Fatty Acids - From Biosynthesis to Human Health

Edited by Erik Froyen

Published in London, United Kingdom

Fatty Acids - From Biosynthesis to Human Health
http://dx.doi.org/10.5772/intechopen.100947
Edited by Erik Froyen

Contributors
Gatot Nyarumenteng Adhipurnawan Winarno, Mathew John, Mohan Vanditha, Sonu Das, Alicia Román-Martínez, Muralidharan Velappan, Deecaraman Munusamy, Qiu-Xing Jiang, Felix Chin, Edgar Antonio Estrella Parra, Erick Nolasco Ontiveros, Maria Elizbeth Alvarez-Sánchez, Rodrigo Arreola, Adriana Montserrat Espinosa González, Ana María García Bores, Roberto Eduardo Lopez-Urrutia, Ignacio Peñalosa Castro, María del Socorro Sánchez Correa, Guangzhen Wu, Qifei Wang, Jin Wang, Grace Chen, Kumiko Johnson, Savina Nodari, Francesco Fioretti, Akio Nakamura, Hikari Nakamura, Ritsuko Kawaharada

First published in London, United Kingdom, 2023 by IntechOpen
IntechOpen is the global imprint of INTECHOPEN LIMITED, registered in England and Wales, registration number: 11086078, 5 Princes Gate Court, London, SW7 2QJ, United Kingdom
Printed in Croatia

British Library Cataloguing-in-Publication Data
A catalogue record for this book is available from the British Library

Additional hard and PDF copies can be obtained from orders@intechopen.com

Fatty Acids - From Biosynthesis to Human Health
Edited by Erik Froyen
p. cm.

This title is part of the Biochemistry Book Series, Volume 42
Topic: Metabolism
Series Editor: Miroslav Blumenberg
Topic Editor: Yannis Karamanos

Print ISBN 978-1-80356-440-1
Online ISBN 978-1-80356-441-8
eBook (PDF) ISBN 978-1-80356-442-5
ISSN 2632-0983

For EU product safety concerns:
IN TECH d.o.o., Prolaz Marije Krucifikse Kozulić 3, 51000 Rijeka, Croatia,
info@intechopen.com or visit our website at intechopen.com.

IntechOpen Book Series
Biochemistry
Volume 42

Aims and Scope of the Series

Biochemistry, the study of chemical transformations occurring within living organisms, impacts all of the life sciences, from molecular crystallography and genetics, to ecology, medicine and population biology. Biochemistry studies macromolecules - proteins, nucleic acids, carbohydrates and lipids –their building blocks, structures, functions and interactions. Much of biochemistry is devoted to enzymes, proteins that catalyze chemical reactions, enzyme structures, mechanisms of action and their roles within cells. Biochemistry also studies small signaling molecules, coenzymes, inhibitors, vitamins and hormones, which play roles in the life process. Biochemical experimentation, besides coopting the methods of classical chemistry, e.g., chromatography, adopted new techniques, e.g., X-ray diffraction, electron microscopy, NMR, radioisotopes, and developed sophisticated microbial genetic tools, e.g., auxotroph mutants and their revertants, fermentation, etc. More recently, biochemistry embraced the 'big data' omics systems. Initial biochemical studies have been exclusively analytic: dissecting, purifying and examining individual components of a biological system; in exemplary words of Efraim Racker, (1913 –1991) "Don't waste clean thinking on dirty enzymes." Today, however, biochemistry is becoming more agglomerative and comprehensive, setting out to integrate and describe fully a particular biological system. The 'big data' metabolomics can define the complement of small molecules, e.g., in a soil or biofilm sample; proteomics can distinguish all the proteins comprising e.g., serum; metagenomics can identify all the genes in a complex environment e.g., the bovine rumen.

This Biochemistry Series will address both the current research on biomolecules, and the emerging trends with great promise.

Meet the Series Editor

Miroslav Blumenberg, Ph.D., was born in Subotica and received his BSc in Belgrade, Yugoslavia. He completed his Ph.D. at MIT in Organic Chemistry; he followed up his Ph.D. with two postdoctoral study periods at Stanford University. Since 1983, he has been a faculty member of the RO Perelman Department of Dermatology, NYU School of Medicine, where he is codirector of a training grant in cutaneous biology. Dr. Blumenberg's research is focused on the epidermis, expression of keratin genes, transcription profiling, keratinocyte differentiation, inflammatory diseases and cancers, and most recently the effects of the microbiome on the skin. He has published more than 100 peer-reviewed research articles and graduated numerous Ph.D. and postdoctoral students.

Meet the Volume Editor

Dr. Froyen is an assistant professor in the Department of Nutrition and Food Science, at California State Polytechnic University, USA. He has a Ph.D. in Nutritional Biology from the University of California, Davis, USA. His research interests include the mechanisms by which fatty acids and phytochemicals decrease the risk factors for cardiovascular disease and cancer. He has given many conference presentations and published many journal articles on these topics. He also teaches courses in nutrition, nutrition research, nutrient metabolism, and nutritional genomics.

Contents

Preface

Interest in fatty acids is growing, as they have been demonstrated to be associated with cardiovascular disease, cancer, inflammation, autoimmune diseases, metabolic disorders, and other diseases and health conditions.

Cardiovascular disease and cancer are top contributors to death worldwide. Increasing public education on nutrition and therapeutic strategies to decrease the risk factors for cardiovascular disease and cancer, as well as approaches to treat inflammation, autoimmune diseases, metabolic disorders, and other diseases, has become imperative. Moreover, the biosynthesis and processing of fatty acids are important for addressing human health and environmental concerns.

Dietary patterns have been suggested to influence the risk factors for developing cardiovascular disease, cancer, and other chronic diseases. Furthermore, individuals with lipid metabolic disorders need unique lipid recommendations to address changes in lipid metabolism. Lipidomics is one such technique to help provide nutrition and therapeutic recommendations for genetic disorders, in addition to recommendations for individuals with certain diseases and health conditions. Synthesizing and processing fatty acids can make them available for individuals experiencing health issues or for those interested in lowering the risk factors for chronic diseases.

This book highlights methods to produce fatty acids as well as provides fatty acid recommendations to decrease the risk factors for chronic diseases. Moreover, it discusses nutrition, fatty acid, and therapeutic strategies for individuals with certain diseases and health conditions.

I hope this book encourages further research on the mechanisms by which fatty acids decrease the risk factors for chronic diseases. In addition, I hope research continues to provide fatty acid recommendations for individuals with certain diseases and health conditions.

Erik Froyen, Ph.D.
Assistant Professor,
Department of Nutrition and Food Science,
California State Polytechnic University,
Pomona, CA, USA

Section 1

Biosynthesis and Processing of Fatty Acids

Chapter 1

Oleochemical Processing Technology: From Process Engineering and Intensification Techniques to Property Models for the Exploitation of Residual Marine Oils

Alicia Román-Martínez

Abstract

A review of the efforts done in process engineering aspects, such as process optimization and process intensification of residual oils processing, are described and discussed. It should be emphasized that the important characteristics of marine oils be determined for a good process design practice, especially, the quality attributes of the residual oil as a raw material. Finally, some property prediction models that have been proposed are indicated. All these aspects: 1) novel process engineering tools, 2) quality characterization, and 3) property models, are important for sustainable products and processes implementation in a circular economy.

Keywords: lipid processing, lipid characterization, lipid property models, marine oils, marine oil processing

1. Introduction

Lipids are an indispensable part of the diet and participate in numerous important biological functions. Products obtained from lipid sources (e.g., animal, vegetable fats, and oils) have gained relevance in the last two decades as the need for healthier food products or the growing interest in biofuels, among others, have increased. Consequently, the processing of fats and oils has been of vital importance throughout the history of humanity [1]; lipids have been used since time immemorial, mainly for food purposes, although there are other recent applications, such as paints, lubricants, personal care products, and (bio)fuels, as well as in animal feed [2]. The increase in the variety of lipid products and population growth have caused the annual demand for fats and oils to practically double since the end of the 1990s worldwide. The total production volume for the 12 most important oils was around 90.5 million metric tons in the year 2000, while it is currently estimated at 207.5 million tons per year [3].

On the other hand, it is known that food production systems in general represent around 20–30% of the consumption of natural resources worldwide [4]. Therefore, it has become essential to consider a change toward sustainable systems at all stages of the design and implementation of processes, which implies that they have the capacity to meet the present needs of society without compromising those of future generations. As in most human activities, in the oleochemical industry, there is a continuous effort to reduce overall production costs and, more recently, to mitigate environmental impact, meet consumer needs and promote social progress [5].

In order to address these challenges, the concept of circular economy (CE) was developed to transform productive activities from a linear perspective to a circular one, increasing efficiency in the use of resources, as shown in **Figure 1** [6, 7]. Optimal use of available resources prolongs their functionality and value, promotes production patterns with cycle closure, and thus reduces waste generation [8]. This implies, among other activities, the recycling and reuse of biomass sources, and organic waste to generate various products and materials, including food, chemicals, biopolymers, fuels, and bioenergy [9, 10].

In this context, numerous proposals have been made to exploit oil/lipid-based feedstocks that could lead to waste utilization and valuable products [11–13]. Among these, waste from the fishing industry has attracted great interest as it is a source of minerals, proteins, and fats with high potential for use [14]. About 30–40% of the fish is consumed fresh, while the rest is processed for marketing and other purposes [15]. However, not all parts of the fish are consumed, as substantial parts (e.g., fish oil) are often wasted. The processing of these marine oils, in addition to offering the possibility of obtaining biofuels, can be used to extract and refine commercial by-products of high nutraceutical value, such as tocopherols, sterols, and omega-3 (ω-3) fatty acids, particularly eicosapentaenoic acid (EPA) and docosahexaenoic acid (DHA) [16].

These polyunsaturated fatty acids (PUFAs) from the omega-3 family have received particular attention due to their unique health benefits. Epidemiological studies conducted in the 1970s indicated a remarkably low incidence of death from ischemic

Figure 1.
General diagram of the life cycle of products with a circular economy approach. Adapted from [6].

heart disease in Eskimos from Greenland, as well as in Japanese and Turkish coastal populations with diets rich in fish constituents [17, 18]. Since then, numerous biochemical and biomedical investigations have been published in relation to the effects of these fatty acids, whose consumption is critical for physical and mental health, and playing an important role in infant brain and eye development [19, 20].

To obtain the benefits of this family of lipids, the World Health Organization (WHO) and the North Atlantic Treaty Organization (NATO) recommend a daily intake of 200 to 500 milligrams of EPA and DHA [21]. According to the UK Scientific Advisory Committee on Nutrition Guidelines (2004), and the Dietary Guidelines Advisory Committee Report (2005) published by the Department of Health and Human Services of the United States, it is recommended to consume at least two servings of fish in the diet per week [22].

However, these recommendations are hardly met. In 2010, the global mean intake of ω-3 fats from marine sources was 163 mg/day, with huge regional variations; only in 45 of the 187 countries, the average consumption was greater than 250mg/day, while in 100 countries there is a very low consumption (<100 mg/day), generally in regions of Africa, Asia, and Latin America, representing the 66.8% of the world adult population [23].

This situation has led to the rapid growth of the ω-3 supplement industry from marine oils. At the beginning of the last decade, only about 5% of the world's fish oil production was destined for the extraction of its ω-3 content for use as ingredients and dietary supplements, while the remaining fraction was discarded or used for fish farming or aquaculture (food for raising fish) [19]. Currently, the consumption of marine oils is expanding due to the recognition and dissemination of their beneficial properties, and this demand is expected to continue growing, with the world market for ω-3 reaching around 6955 million dollars for the year 2022 [24]. Therefore, the recovery and transfer of these important nutrients from the sea to the human food chain, within the circular economy approach, is a relevant opportunity to promote economic growth, environmental protection, and human health in general.

In addition to the increase in demand, progress in the investigation of methodologies for the extraction, purification, and stabilization of ω-3 fish oil has made significant progress in recent decades, although its application has not been widely extended on a large scale in the industry due to the challenge it represents. Marine oils are complex mixtures of fatty acids with variable chain lengths and degrees of unsaturation, which makes their separation difficult [25]; in addition to being highly susceptible to deterioration processes, such as thermal oxidation and enzymatic hydrolysis [26].

Additionally, unlike the chemical industry, the development of the oleochemical industry has not had such rapid progress in terms of modeling thermo-physical properties and development of adequate computational tools for the design, analysis, and optimization of processes related to lipids, coupled with the lack of integration of sustainable aspects [27–29]. Marine oil processing involves a variety of stages, from the extraction and recovery of crude oil to the refinement and modification of the final product. Typical unit operations include fluid transport, heat transfer, and separation processes, such as adsorption, phase separation, crystallization, filtration, chemical synthesis (interesterification, hydrogenation), and vacuum steam distillation, among others [25, 30].

Within this context, the decision-making set when designing a plant, also called "process synthesis," must balance the criteria of sustainability and incorporate the circular economy within the design, for which different tools and technologies have been developed. Among them, process intensification (PI) may be a promising engineering approach. PI is considered as any novel equipment, processing technique, or method that, compared to conventional ones, offers a substantial improvement in the efficiency and/or performance of manufacturing (bio)chemical products [31]. Another complementary definition suggests an optimization of the basic functional principles of chemical processes (momentum, heat, and mass transfer) in the design of operations [32].

A wide range of applications and developments in the oleochemical industry can be considered intensified, as shown in **Table 1**. For example, nano-neutralization seeks to integrate degumming and neutralization operations to reduce the number and size of equipment, as well as the dosage of necessary reagents and the generation of waste [34]. Companies, such as Alfa Laval® or Unilever®, have developed strategies to improve the configuration of the process and the operating parameters in order to optimize the removal of impurities depending on the nature and quality of the oil [33]. Microwave irradiation has been used to improve the efficiency of biodiesel production, reducing reaction time with energy savings of up to 44% [35]. Similarly, ultrasound-assisted transesterification of various vegetable oils has proven to be effective by increasing the contact surface between the alcohol/oil phases, improving mass transfer and reducing the amount of catalyst needed for biodiesel production [36, 54], as well as ethyl esters of fish oil [37].

PI principle applied	PI Technology	Applications in the oleochemical industry	References
Integration of unit operations. Hybrid separations	Special Combi-Mix Alfa Laval® degumming	Degumming and neutralization	[33]
	Nano-neutralization	Degumming and neutralization	[34]
Higher performance for a given process or equipment size. Improvement in the transfer of heat, mass or momentum through alternative energy sources	Microwave/Ultrasound-assisted reactors	Transesterification reactions for the production of biodiesel or ethyl esters	[35–37]
	Membrane technologies	Degumming, neutralization, deodorization, fatty acid concentrate	[38, 39]
	Technology with supercritical fluid (CO_2)	Oil extraction, fatty acid concentrate	[40–43]
Smaller volume of equipment for a given process or output. Improved heat, mass, or momentum transfer	Molecular distillation	Deodorization, fatty acid concentrate, recovery of valuable lipid compounds, such as tocopherols	[44–46]
	Chromatographic methods	Fatty acid concentrate	[47–49]
Less use of ancillary services and/or raw material flows (reagents, solvents, electricity...). Greater yields	Trysil® Silica Adsorbent Hydrogels	Oil washing and bleaching	[50]
	Enzymatic alternatives	Oil extraction, degumming, neutralization, lipid modification (trans/interesterification)	[51–53]

Table 1.
Some process intensification (PI) options with reported applications in the oleochemical industry.

Other strategies include oil extraction or fatty acid separation using supercritical fluids [40], membrane technology for impurity filtration [38], or PUFA concentrate [39, 55], in addition to enzymatic alternatives for different reactions [51, 52].

Evaluating different process alternatives (conventional or intensified) from a sustainable perspective normally requires multi-objective or multi-criteria decision-making that must often consider antagonistic criteria, that is, when deciding to prefer one, the other criteria are affected; consequently, you are forced to make more informed and, therefore, better decisions. In these problems, there is not usually a single optimal solution and it is necessary to use a decision-making methodology to be able to differentiate the alternatives and prioritize the possible solutions.

To achieve this goal, the life cycle analysis (LCA) can be a very helpful tool, since it involves metrics that characterize the environmental impact of a product or process in all its stages, from the extraction of the raw material to the final disposal of waste [56], which obviously allows integrating the concept of circular economy within the process design [57]. There are several studies that demonstrate the potential application of LCA in oleochemical systems [58]. Additionally, the use of mathematical optimization methodologies allows evaluating these multi-criteria indicators in a robust way, as well as implementing of this set of strategies in software or process simulators [59, 60] in order to speed up the decision-making process.

Table 2 presents a synthesis of different studies carried out on the design of processes involving lipids, and the tools or approaches addressed in each one. It is possible to observe that there is rarely a comprehensive vision of the strategies or indicators that would allow achieving an optimal sustainable process (e.g., there are few studies that address the evaluation of social metrics, a pillar of sustainability as important as the economic or the environmental). Therefore, the proposal of systematic methodologies that involve the mathematical evaluation of alternatives through a balance of sustainable indicators is vital for the design of processes in the context of the circular economy. The integration of these approaches together with social aspects, such as risk analysis and the opportunity to mitigate the nutritional deficiencies of vulnerable communities (by valorizing the sources of ω-3), complement a promising strategy that would allow the generation and evaluation of robust alternatives of processing aimed at the development of a more sustainable industry.

2. Marine oil processing

The processing of plant seeds or animal tissues into edible oils can be divided into four sets of operations: extraction, refining, conversion or modification, and stabilization [77, 78]. Oil extraction consists of pressing the material or matrix to separate the crude oil from the protein-rich solids or moderately pressing it with solvent, usually hexane. The defatted solids are known as cake or flour. The oil obtained is considered “crude” because it contains undesirable components, such as pigments, phospholipids, free fatty acids, and unpleasant flavors and odors, so it must be refined to remove these contaminants, and thus obtain a high-quality edible oil. Refined oils consist mainly of triacylglycerides (>98%) that can be subsequently modified through processes, such as hydrogenation, winterization, fractional crystallization, or interesterification, whose objective is to obtain properties different from the original oil, such as the conversion of liquid oil into semi-solid fats or the fatty acid concentrate of interest (e.g., ω-3). The stabilization of the final product to suit the needs of the consumer depends on the use that it will have. Plasticizing and tempering are

Characteristics and references	Approach or tool													
	Process design/ simulation	Process intensification (PI)	Sustainability					Optimization						
			Economic analysis	Environmental impact	Social indicators	Circular economy	Life Cycle Assessment (LCA)	Mono-objective	Multi-objective	Linear Programming (LP) or Non-Linear Programming (NLP)	Mixed-Integer Non-Linear Programming (MINLP)	Superstructure Approach	Stochastic (uncertainties)	Sensitivity analysis
Biodiesel production from palm and rapeseed oils, compared with petroleum-based diesel [61]	•		•	•			•							
Comparison and impact of different methods to produce biodiesel from rapeseed oil [62]	•		•	•	•									
Risk analysis in the design of edible oil refineries [63]	•				•									
Comparison of three alternatives for biodiesel production from discarded oils [64]	•	•		•		•		•						
Biodiesel production from fresh and residual vegetable oils, using a conventional and an intensified method [65]	•	•	•			•								•
Analysis of biodiesel production from oil discarded in a restaurant [66]	•		•	•		•								
Life Cycle Analysis for the production of five vegetable oils [67]				•			•						•	
Design of the solvent-based butter fractionation process using a systematic approach [11]	•		•	•			•							

Characteristics and references	Approach or tool													
	Process design/ simulation	Process intensification (PI)	Sustainability					Optimization						
			Economic analysis	Environmental impact	Social indicators	Circular economy	Life Cycle Assessment (LCA)	Mono-objective	Multi-objective	Linear Programming (LP) or Non-Linear Programming (NLP)	Mixed-Integer Non-Linear Programming (MINLP)	Superstructure Approach	Stochastic (uncertainties)	Sensitivity analysis
Transesterification and separation of oleic acid by reactive distillation [68]	•	•												
Modeling of biodiesel production in a continuous tubular reactor with static mixer [69]	•	•	•											•
Production of glycerol and biodiesel from algal oil [70]	•		•					•		•				•
Systematic analysis of the soybean oil extraction process. Prediction of thermodynamic properties [71]	•							•		•				
Hybrid optimization model for the design of a biorefinery using a sustainable perspective [72]		•	•						•	•		•		
Optimization of a glycerol purification process and solvent crystallization of a vegetable oil [73]	•		•	•				•			•	•	•	
Simulation of a palm oil extraction, refining and fatty acid separation process [74]	•		•	•					•		•	•	•	
Analysis of a biodiesel production process from discarded fish oil [12]			•			•								

Characteristics and references	Approach or tool													
	Process design/ simulation	Process intensification (PI)	Sustainability					Optimization						
			Economic analysis	Environmental impact	Social indicators	Circular economy	Life Cycle Assessment (LCA)	Mono-objective	Multi-objective	Linear Programming (LP) or Non-Linear Programming (NLP)	Mixed-Integer Non-Linear Programming (MINLP)	Superstructure Approach	Stochastic (uncertainties)	Sensitivity analysis
Design of a biorefinery for the production of Omega-3 concentrates from discarded fish oil [42]	•	•	•	•		•								
Comparison of the production of Omega-3, protein and biofuel from algal and fish oil [75].	•			•			•							•
Economic evaluation and simulation of an enzymatic process for the production of biodiesel from rapeseed oil [76]	•	•	•											

Table 2.
Examples of different approaches and tools used in lipid process design.

operations designed to stabilize mixtures used in margarines and shortenings, while nutraceutical oils used as supplements usually undergo encapsulation or microencapsulation processes to prevent deterioration and facilitate their consumption or incorporation into other food products [79].

2.1 Extraction methods

The method chosen for oil extraction depends on the nature of the raw material, as well as the capacity of the industrial plant [80]. Particularly, for fish oil, it is obtained in most cases through the "wet-rendering" process, which consists of heating or "cooking" fresh fish with steam to break the cellular structure, and then pressing to separate the particles, liquid, and solid fractions. The solid fraction is dried to produce fishmeal.

The liquid fraction contains water, oil, and minor amounts of suspended solid material. This liquid is processed using centrifuges and decanters to remove solids and separate the water from the oil, which is stored for later use as crude oil. The water fraction is evaporated and recirculated to the solid fraction before drying [81].

Other traditional methods for its extraction include enzymatic hydrolysis, autolysis, dry rendering, solvent (hexane) extraction, and more recently innovative and ecologically friendly methods, such as supercritical fluid extraction, enzyme extraction, and extraction using ultrasound or microwaves have been used [40].

2.1.1 Storage and preservation of crude oil

It can be considered that the processing of oil starts from the storage tanks since the quality and performance are affected by the conditions in which it is stored. In particular, it is desired to avoid increases in free fatty acids, oxidation products, color changes, and contamination by insoluble impurities.

In general, tanks should store oil at room temperature ($<25°C$), or cooler when practical and possible. It is recommended to use containers made of stainless steel that presents the least possibility of iron contamination, for example, coated food-grade stainless steel (316). It is critical to have completely sealed tanks, as well as to fill the upper space with nitrogen to displace the air and avoid its contact with the oil since PUFAs in particular are prone to oxidation phenomena, as will be described later [26, 82].

2.2 Oil refining

Refining generally refers to the removal of non-triacylglycerol fatty materials, as well as pretreatment and deacidification or neutralization operations. Consumers typically want mild or neutral-flavored, light-colored, and oxidatively stable oils, so the goal is to remove impurities that may cause the product to have an off-color or off-flavor or cause harmful metabolic effects for human consumption.

Edible oils can be refined through chemical or physical processes. Each method has its own specific advantages, and the use of one or the other depends on quite a bit on the quality and type of oil involved. In chemical or alkaline refining, caustic soda is used to neutralize the free fatty acids in the oil, while in physical refining this stage is skipped, removing these acids (to a lesser extent) by single-stage distillation during the process of deodorization. In general, when the acidity of the oil exceeds approximately 2%, chemical refining is carried out, while for lower values, physical refining is

Method	Characteristics	Applications	Limitations
Refining or physical deacidification	It uses vacuum steam distillation, which removes free fatty acids and aromatic compounds. Requires high temperatures and depends on efficient prior degumming.	Suitable for oils with a high content of free fatty acids. Low capital and operating cost. Higher oil yield. Reduced amount of effluents (does not produce soaps).	Pretreatments are more rigorous. Not suitable for heat-sensitive oils (cotton, fish). Possible thermal polymerization. Controlled rate of elimination of free fatty acids.
Refining and chemical acidification	Industrially, it is the most common method. It is done by adding an alkaline substance (caustic soda) that precipitates free fatty acids as soaps.	Versatile (produces good quality oil from any type of source). Multiple effects; purified, degummed, neutralized and partially bleached. Chemically more stable product.	Excessive loss of neutral oil when there is a high content of free fatty acids. Produces waste (soaps). Oil hydrolysis.
Mixed or miscellaneous refining	More suitable for the integration of an extraction and refining plant, since it combines both processes, mixing a solvent (hexane) that helps to separate impurities during degumming and neutralization.	Lower strength of caustic solution. Higher separation efficiency. Minimal oil occlusion in soap scum. Better color in the final product. Water washout removed.	Higher investment costs. Loss of solvent (requires careful operation and increased maintenance). For efficient operation, the oil concentration in the micelle should be ~50% (two-stage solvent removal).

Table 3.
Industrial oil refining methods. Adapted from [33, 80, 83].

carried out, provided that the quality of the oil allows it [83]. **Table 3** presents the main characteristics, advantages, and disadvantages of these refining approaches.

Regardless of the approach, refining is a set of operations designed in such a way as to minimize the loss of “neutral” oil and maximize the availability of beneficial components. The traditional process [33] includes the stages of degumming, neutralization, bleaching, deodorization, and in some cases, winterization (although the latter is considered a PUFA concentrate method). Each stage is important to remove the different impurities [84].

Evolution in the ω-3 fatty acid market has led to the development of new processes for treating marine oils, designed to preserve these acids, while reducing impurities. In some cases, these processes were initially expensive, but due to the evolution of this market, they are now considered conventional processes [16]. The processing steps and the compounds removed by them can be summarized in **Table 4**.

2.3 PUFA modification and concentrate

Oil modification processes involve a substantial change in their physical behavior and structural properties, unlike the previously discussed refining processes where their effect is aimed at removing impurities and improving organoleptic properties, as well as nutritional value [1]. Chemical modifications offer the possibility of changing the properties of fats and oils within wide ranges, making them suitable for many uses.

Technique or Stage	Purpose
Storage	Insoluble impurities, residual moisture, and some phospholipids precipitate in the tanks.
Degumming	Remove phospholipids, sugars, resins, protein compounds, trace metals, and other materials.
Alkaline refining (Neutralization)	Remove free fatty acids, pigments, phospholipids, water-soluble and insoluble material, and trace metals.
Silica washes/treatment	Soaps are produced during the neutralization stage, oxidation products, and trace metals.
Dying	Moisture reduction.
Adsorbent bleaching and/or carbon treatment	Pigments, oxidation products, trace metals, sulfur compounds, dioxins, furans, and polycyclic aromatic hydrocarbons.
Winterization	High melting point triacylglycerides and waxes. Used to increase unsaturated triacylglycerides.
Deodorization	Eliminates volatile compounds: free fatty acids, mono-diacylglycerides, aldehydes, ketones, chlorinated hydrocarbons, and pigment decomposition products, as well as oxidation. It is usually the final step and results in a mild-tasting oil.
Vacuum extraction, molecular or short distillation.	Removes chlorinated hydrocarbons, fatty acids, PCB oxidation products, and free cholesterol. It can be used to replace deodorization.

Table 4.
Stages of oil processing and the main impurities removed in each one [16].

Two main modification technologies are generally considered in the edible oil industry:

i. Hydrogenation or hardening: It consists of the addition of protons (hydrogen atoms) to the unsaturated sites of the hydrocarbon chain of fatty acids, transforming these into saturated bonds. It was originally used to improve the stability of oils with large amounts of PUFAs, such as marine oils, however, the use of this process has been considerably reduced as it has been identified as an important source not only of saturated fats but also of trans fatty acids in the human diet, associated with an increased risk of heart disease [85]. It should also be considered that, by eliminating unsaturation, its nutritional value is lost [1].

ii. Esterification reactions: Two terms are commonly used in the literature to designate a wide variety of ester-ester exchange reaction mechanisms [52]:

   a. Interesterification: General term for reactions between an ester and fatty acid, alcohol, or other esters, although it can only describe ester-ester exchanges or randomization of oils and fats by alkaline or enzymatic catalysis.

   b. Transesterification: Term used for ester-ester exchanges, including acidolysis and alcoholysis. It is mostly used to describe reactions of oils with methanol or ethanol for the production of biodiesel or ethyl esters.

Technique	Description
Interesterification	Re-arrangement of the triacylglycerides in the oil in a particular or more random distribution.
Hydrolysis or esterification	Separation of fatty acids from triacylglycerol molecules, producing free fatty acids or esters with glycerin as a by-product.
Urea complex	By dissolving the fatty acids or esters in methanol or ethanol, adding urea, and reducing the temperature, it is possible to precipitate a complex that traps the saturated and monounsaturated ones, resulting in a PUFA concentrate. Depending on the type of oil, the losses can be significantly large.
Winterization / Dry fractionation	Separation of high melting point (saturated) waxes and triacylglycerides by cooling the oil below their crystallization temperature, and their subsequent separation by filtration.
Molecular distillation	This stage can be used to remove free cholesterol or further concentrate esters or fatty acids.
Extraction with supercritical fluids (SCF)	Purification to obtain >85% pure compounds and remove cholesterol, using fluids at supercritical conditions of temperature and pressure.
Preparative High-Performance Liquid Chromatography (HPLC)	Purification of esters or fatty acids to produce compounds with more than 95% purity. It is a chromatography-based technique and can be very expensive and complicated.
Re-esterification	Conversion of purified fatty acids or esters back to their triacylglycerol form, which is considered more natural and more bioavailable to the body. It is made by a catalytic reaction with excess glycerol.

Table 5.
Techniques for the concentration and purification of omega-3 fatty acids. Adapted from [16, 86].

Once their properties have been modified, there are additional steps that can be used to convert refined marine oils into concentrates and relatively purified esters or fatty acids into ω-3 fractions (**Table 5**).

Many of these processes described above can be adapted, combined, or in some cases improved through the use of enzymes [25, 52]. The use of simple processes based on green chemistry, low pressures, and methods that avoid toxic solvents and high temperatures have been positioned to replace conventional methods for the extraction, purification, and stabilization of marine oils rich in ω-3 fatty acids [19].

3. Analysis of oils and quality parameters

Regardless of where they come from, oils and fats are similar in terms of the effects of the aforementioned impurities on processing and final product quality. Therefore, it is necessary to determine certain key analytical values to be able to modify or establish the process conditions in order to obtain a product of satisfactory quality [1]. Each analysis provides specific information to the manufacturer, as well as to future consumers.

3.1 Analytical characterization

Tests that are important to consider in determining the quality and handling of crude fish oil (or any oil) are as follows:

i. Sampling methodology: The objective of sampling is to obtain a manageable quantity of oil whose properties correspond as closely as possible to those of the original source, as well as to ensure its transport and storage. The considerations according to ISO 5555:2001 [87] must be taken into account.

ii. Determination of percentage of free fatty acids (FFA): Free fatty acids appear with the decomposition of lipids, being indicators of degradation. Their percentage must remain below the standards established for human consumption, so they must be neutralized during the refining of the oil. For its determination, the AOCS Ca 5a-40 methodology [88] is used.

iii. Determination of peroxide index (PI): It is the amount (expressed in milliequivalents of active oxygen per kilogram of fat) that indicates the initial oxidation state of the oil. A maximum value for refined oils of 5 to 10 meq/kg of O2 should be considered. The methodology used is AOCS CD 8b-90 [89].

iv. Determination of the p-anisidine index (AI): The use of the peroxide index is limited to the initial stages of lipid oxidation, so the anisidine index is established as a useful measure of secondary oxidation products to evaluate the past of the oil and predict its stability. Its value must be less than 20meq/kg of O2. The methodology used is AOCS Cd 18–90 [90].

v. Determination of TOTOX (total oxidation) index: Indicates the general state of oxidation of the oil, the lower it is, the better its quality. It is simply calculated by combining the peroxide and p-anisidine values with the formula AI+2PI. It must be less than 26 meq/kg of O2.

vi. Determination of iodine index (II): This is a measure of the unsaturation of fats and oils expressed in terms of the percentage of iodine absorbed per gram of sample. The higher this index, the greater the number of double bonds. Its value can range from 90 to 150 for vegetable oils, and greater than 150 for fish oils. It is determined by the AOCS method number Cd 1d-92 [91].

vii. Determination of the fatty acid profile: It is carried out to know the composition of the fatty acids that make up the triacylglycerides of the oil; in this way, it is possible to know the amount of saturated, monounsaturated, and polyunsaturated present. It is carried out by gas chromatography of the previously methylated sample to convert the triacylglycerides into fatty acid methyl esters. For marine oils, the AOCS Ce 1b-89 methodology is used [92].

3.2 Composition of marine oils

Particularly for fish oils, the constitution and proportion of its lipid components depend on the extraction process, the species, and the geographical and environmental conditions; hence, the emphasis on carrying out an adequate characterization that allows obtaining information about the composition of fatty acids and their physicochemical properties, which can give an idea of the quality of the oil and its nutritional value, important considerations for a subsequent design of oil processing according to the desired product.

For this type of oil, triacylglycerides occupy the largest proportion (~90%); the moisture content and the percentage of free fatty acids vary according to the extraction treatment and storage conditions, these being around 0.5–1% and 2–5%, respectively [81]. The amount of phospholipids can vary from 1 to 1.5%, and in terms of the content of the unsaponifiable fraction (sterols, tocopherols, pigments, alcohols, and hydrocarbons), it is normally less than 2%, although it can constitute up to 8% of the oil under certain seasonal and feeding conditions [93].

The fatty acid profile allows a chromatogram to be obtained from which the type and content of fatty acids belonging to saturated, monounsaturated fatty acids or PUFAs is determined. The profile of fish oils is known to be complex, varying by marine species. There are fatty acids that occur in trace amounts and some are generally common to all species. For example, high contents of PUFAs have been found in tuna oil (35% w/w) [94], in cod liver (18% w/w), and anchovies (16.5% w/w) [95].

Fatty acid		**Anchovy**	**Tuna**	**Krill**	**Salmon**		**Cod liver**
					Wild	**Farm**	
Myristic	C14:0	2.7–11.5	ND-5.0	5.0–13.0	2.0–5.0	1.5–5.5	2.0–6.0
Pentadecanoic	C15:0	ND-1.5	ND-2.0	NA	ND-1.0	ND-0.5	ND-0.5
Palmitic	C16:0	13.0–22.0	14.0–24.0	17.0–24.6	10.0–16.0	6.5–12.0	7.0–14.0
Palmitoleic	C16:1	4.0–12.6	ND-12.5	2.5–9.0	4.0–6.0	2.0–5.0	4.5–11.5
Margaric	C17:0	ND-2.0	ND-3.0	NA	ND-1.0	ND-0.5	NA
Stearic	C18:0	1.0–7.0	ND-7.5	NA	2.0–5.0	2.0–5.0	1.0–4.0
Vaccenic	C18:1 ω7	1.7–3.7	ND- 7.0	4.7–8.1	1.5–2.5	NA	2.0–7.0
Oleic	C18:1 ω9	3.6–17.0	10.0–25.0	6.0–14.5	8.0–16.0	30.0–47.0	12.0–21.0
Linoleic	C18:2 ω6	ND-3.5	ND-3.0	ND-3.0	1.5–2.5	8.0–15.0	0.5–3.0
α-Linolenic	C18:3 ω3	ND-7.0	ND-2.0	0.1–4.7	ND-2.0	3.0–6.0	ND-2.0
γ-Linolenic	C18:3 ω6	ND-5.0	ND-4.0	NA	ND-2.0	ND-0.5	NA
Stearidonic	C18:4	ND-5.0	ND-2.0	1.0–8.1	1.0–4.0	0.5–1.5	0.5–4.5
Arachidic	C20:0	ND-1.8	ND-2.5	NA	ND-0.5	0.1–0.5	NA
Eicosenoic ω-9	C20:1 ω9	ND-4.0	ND-2.5	NA	2.0–10.0	1.5–7.0	5.0–17.0
Eicosenoic ω-11	C20:1 ω11	ND-4.0	ND-3.0	NA	NA	NA	1.0–5.5
Arachidonic	C20:4 ω3	ND-2.5	ND-3.0	NA	0.5–2.5	ND-1.2	ND-1.5
Eicosatetraenoic	C20:4 ω3	ND-2.0	ND-1.0	NA	1.0–3.0	0.5–1.0	ND-2.0
Eicosapentaenoic	C20:5 ω3	5.0–26.0	2.5–9.0	14.3–28.0	6.5–11.5	2.0–6.0	7.0–16.0
Heneicosapentaenoic	C21:5 ω3	ND-4.0	ND-1.0	NA	ND-4.0	NA	ND-1.5
Erucic	C22:1 ω9	ND-2.3	ND-2.0	ND-1.5	ND-1.5	3.0–7.0	ND-1.5
Ketoleic	C22:1 ω11	ND-5.6	ND-1.0	NA	1.0–1.5	NA	5.0–12.0
Docosapentaenoic	C22:5 ω3	ND-4.0	ND-3.0	ND-0.7	1.5–3.0	1.0–2.5	0.5–3.0
Docosahexaenoic	C22:6 ω3	4.0–26.5	21.0–42.5	7.1–15.7	6.0–14.0	3.0–10.0	6.0–18.0

Table 6.
Ranges of fatty acid profiles of some fish oils determined by gas-liquid chromatography (expressed as percentages of total fatty acids). ND = not detected, defined as <0.05%. NA = not available; ω3 = omega-3 series, ω6 = omega-6 series, ω7 = omega-7 series, ω9 = omega-9 series [95].

While the monounsaturated fatty acids, oleic acid (C18:1) is the most abundant, and as for the saturated ones, large amounts of myristic acid (C14:0) and palmitic acid (C16:0) can be found in almost all the species analyzed [94].

It is important to mention that the positional distribution of the different types of fatty acids in the glycerol molecule of triacylglycerides, especially, the ω-3 PUFAs, has a strong influence on digestion and absorption in the human body, as well as on its oxidative stability [96]. In marine oils, DHA is generally found in the sn-2 position, while EPA can be found in any position of the glycerol molecule [97]. **Table 6** below presents a comparison of the fatty acid profile of oils from different fish species reported by the CODEX Alimentarius Commission of the World Health Organization.

Regarding specific tuna oils, various authors in different locations have reported relatively similar fatty acid profiles. The percentages of PUFAs vary from 28 to 44%, with DHA being the most abundant, followed by EPA and, to a much lesser extent, DPA [94, 95, 98]. Oleic acid represents the highest percentage of monounsaturated, and of the saturated, myristic acid stands out.

3.3 Lipid deterioration

Fats and oils can undergo different deterioration processes that, in addition to reducing the nutritional value of the food, produce volatile compounds that impart unpleasant odors and flavors. The degree of deterioration depends on the type of fat or oil, the handling given to it [99], and the storage conditions, being the most affected oils of marine origin, followed by vegetable oils and finally animal fats, due to the fact that the first they contain a greater amount of PUFAs prone to the deterioration [1]. Some examples are:

i. *Lipolysis* - This type of deterioration is due to chemical and/or enzymatic hydrolysis of acylglycerides, breaking the ester bonds of lipids, produced by enzymatic action (lipases) or by heating in the presence of water, and results in the release of fatty acids (now FFA) in the oil.

ii. *Oxidation processes* - The oxidation of lipids is one of the main causes of the deterioration of the organoleptic quality of foods, producing abnormal odors and flavors, generally called "rancid." It is carried out mainly in unsaturated fatty acids, although it is also carried out with other substances of biological interest such as tocopherols and carotenoids [26]. Depending on the factor that acts as a source of energy to start oxidation, it can be classified as photo-oxidation (due to light), thermo-oxidation (when exposed to temperatures greater than 60°C), or auto-oxidation (occurs when organic compounds react with molecular oxygen) [100].

Fats have a certain resistance to oxidation processes depending on the initial natural content of antioxidant and/or prooxidant substances. Oxidation is slow until this resistance is overcome, at which point it accelerates rapidly. To delay these processes, it is possible to add natural or synthetic antioxidants to the oil, which help reduce oxidation through different mechanisms: inactivation of pro-oxidant metals through chelation, elimination of free radicals by phenolic compounds (tocopherols, BHT, and BHA), or removal of simple O2 species by compounds, such as astaxanthin [101].

3.4 Quality requirements

The quality of oil can be defined based on parameters established by various international organizations or regulations, as well as by the use that will be given to the oil. In the case of edible oils destined for concentrated supplements of ω-3, it is suggested to take into account the parameters described below in **Table 7**, in order to ensure a safe product for the consumer, as well as to establish restrictions and conditions that may be included in the time of process design.

4. Models of thermodynamic and physicochemical properties of lipids

As has already been described, vegetable and animal oils are mixtures of lipids, mainly fatty acids in the form of triacylglycerides, phospholipids, and unsaponifiable matter, among others. Therefore, the composition and proportion of these compounds vary significantly according to the source of origin, having a direct impact on the

Characteristic	Disadvantage or potential problem	Specification
Color	Dark oils may contain contaminants normally removed during refining. The color may be indicative of overheating.	No specification
Acidity index	High values of acidity can indicate a low-quality oil or deteriorated during storage. The acidity value is defined as twice the content of free fatty acids.	2 mg KOH/g of refined oil or < 1% FFA max.
Peroxide index (PI)	It is the first measure of oxidation in an oil.	5 meq/g O2 max.
p-anisidine index (AI)	Measures oxidation products; it reflects the oxidation that has taken place during the lifetime of the oil.	20meq/g O_2 max.
TOTOX index	Relationship between the two previous indices reflects the total oxidation of the oil.	26 meq/g O_2 max.
Humidity	Considered an impurity, high levels of humidity during storage can deteriorate the oil.	0.20% max.
Phospholipids	They act as prooxidants and form deposits at the bottom of storage tanks.	<50 ppm phosphorus.
Soaps	They are formed in the presence of moisture and during neutralization, they can reduce the stability of the oil.	0.005% max.
Total cholesterol	Part of the unsaponifiable fraction. It is not usually removed.	No specification
Iron	Considered a pro-oxidant, it is removed during degumming and refining.	1.5 mg/kg in refined oil max.
Copper	Considered a prooxidant, it is removed during degumming and refining.	0.1 mg/kg in refined oil max.
Arsenic	Heavy metal is present in seawater. It is removed during refining.	0.1 mg/kg max.
Lead	Heavy metal is removed during refining.	0.1 mg/kg max.
Mercury	Heavy metal is removed during refining.	0.1 mg/kg max.

Table 7.
Guide to quality parameters and potential areas of problems or disadvantages. Adapted from [81, 95, 102].

Physical characteristics of marine oils	
Specific heat, cal/g	0.50–0.55
Heat of fusion, cal/g	~54
Combustion heat, cal/g	~9500
Melting point, °C	10–15
Flash point, °C as triglycerides as fatty acids	~360 ~220
Boiling point, °C	>250
Specific gravity at 15°C 30°C 45 °C	0.92 0.91 0.90
Viscosity, cp, at 20 °C 50 °C 90 °C	60–90 20–30 ~10

Table 8.
Main characteristics of marine oils in general [93].

physicochemical properties and the appropriate design of the processing applied to the oil to obtain products suitable for use and consumption by humans [81].

Lipids are not commonly tabulated in property databases, and their polyfunctional structure requires a careful model analysis [103]. Properties, such as melting and boiling temperature, density, viscosity, and liquid-vapor equilibrium data, have been experimentally reported for different sources of oils and fats, the most studied being those of vegetable origin. However, the wide range of lipid species and the enormous number of different configurations that fatty acids can have in the triacylglyceride molecule make it difficult to fully characterize all properties of a given species. It is possible to take some general characteristics (**Table 8**) for a preliminary process design.

As not, all the properties of the lipids of interest have been determined in the literature and due to the difficulty, cost and time consumption involved in carrying out complete experimental studies, it is possible to resort to property prediction methods using mathematical models. In particular, group contribution (GC) methods have been widely used to complement the information available for process design [104].

4.1 Group contribution methods

The use of GC methods is based on the structural information of the molecule to predict specific properties; each substance is considered to be formed by the union of suitably defined structural groups, and each one of them is assigned a certain value as a contribution to said property, assuming that it is the same in every compound where it is present. The property of the pure compound is a function of the frequency in which these groups appear and their contributions. In principle, they are additive methods, where the contributions of each group toward property are added to obtain the approximate final value of the property [104–106].

Various prediction methods with GC for physical and thermodynamic properties with a focus on pure lipid compounds have been developed by various authors, among which those reported in **Table 9** stand out.

GC Method	Properties to estimate	Characteristics	Reference
Lydersen	Critical properties: T_c, P_c y V_c	Prototype and predecessor of other more sophisticated models.	[107]
Ambrose	Critical properties: T_c, P_c y V_c	Improvement in the prediction accuracy of critical properties with respect to Lydersen.	[108]
Klincewicz & Reid	Critical properties: T_c, P_c y V_c	Precision is comparable to the ambrose method.	[109]
Chein-Hsiun Tu	Critical temperature T_c	Refined method to obtain the critical temperature of organic compounds.	[110]
Joback & Reid	T_c, P_c, V_c, T_b, T_m, ΔH°_f, ΔG°_f, C°_p, ΔH_v, ΔH_m, ρ_L	Develop models for the prediction of 11 properties of pure compounds.	[111]
Constantinou & Gani (CG)	T_c, P_c, V_c, T_b, T_m, ΔH°_f, ΔG°_f, ΔH_v	More sophisticated model based on a multilevel approach.	[112]
Marrero & Gani (MG)	T_c, P_c, V_c, T_b, T_m, ΔH°_f, ΔG°_f, ΔH_v, ΔH_m	Multilevel approach based on groups of the first, second and third order. It is one of the most widely used today.	[113]
Kolská (2005)	ΔH_v, ΔS_v	Method used for organic compounds in general.	[114]
Kolská (2008)	Heat capacity C°_p	Method used for organic compounds in general based on groups of first, second, and third order.	[115]
Ceriani (2009)	C°_p y ΔH_v	Focus on lipid compounds: fatty acids, esters and fatty alcohols, and TAGs.	[116]
Ceriani (2011)	Liquid viscosity μ_L	Focus on fatty compounds and esters for biodiesel.	[117]
Ceriani (2013)	P^{vp} y ΔH_v	Focus on lipid compounds: fatty acids, esters, fatty alcohols, and TAGs.	[118]
Díaz-Tovar (2011)	Liquid density ρ_L	Review of different GC methods, special emphasis on density and temperature-dependent properties.	[28]
Zeberg & Stenby	T_m y ΔH_m	Method used only for saturated triglycerides.	[119]
Moorthy	T_m y ΔH_m	Method used for saturated and unsaturated TAGs. It is quite accurate, although it requires complex calculations.	[120]
UNIFAC original	Vapor-liquid equilibrium (VLE)	Original, revised, and extended UNIFAC method for EVL prediction.	[121]
LLE-UNIFAC	Liquid-liquid equilibrium (LLE)	UNIFAC method for prediction of ELL.	[122]
Dortmund-UNIFAC	Vapor-liquid equilibrium (VLE)	Modified Dortmund-UNIFAC method for a more accurate prediction of EVL.	[123]

Nomenclature: T_c: critical temperature, P_c: critical pressure, V_c: critical volume, T_b: boiling temperature, T_m: melting temperature, ΔH°_f: standard enthalpy of formation (298 K), ΔG°_f: standard Gibbs free energy (298 K), C°_p: heat capacity, ΔH_v: enthalpy of vaporization, ΔS_v: entropy of vaporization, ΔH_m: enthalpy of fusion, ρ_L: liquid density μ_L: liquid viscosity, P^{vp}: vapor pressure.

Table 9.
Examples of group contribution methods used in lipids and organic compounds in general.

The range of application and the reliability of these methods depend on several factors [124]:

i. The definition of the groups is used to represent the molecular structure of the pure components.

ii. The property prediction model.

iii. The quantity and quality (in terms of information) of the experimental data set used in the regression to estimate the model parameters.

Another challenge with GC methods is that the selected property model may not have all the necessary parameters, such as groups and/or their contributions, for a specific property, so it is necessary, for the construction of the database of properties, to evaluate these factors when selecting the prediction methods to use. It is important to take into account that, as the number of carbon atoms and unsaturation in the molecule increases, the properties reported in the literature are scarce, so, although it is essential to have predictive models for these cases, the accuracy in the estimation of the properties can decrease, due to the sensitivity that these properties have with the exact conformation of the molecule [104].

5. Conclusions

The revaluation of waste and by-products, mainly in the food and oleochemical industries, is a fundamental part of the necessary transition toward a more sustainable and circular economic model. The future development of the industry must not neglect social issues and must focus on providing quality and accessible products to the communities that cooperatively represent the workforce and progress of humanity. In conclusion, the revaluation of food waste (fish oil) in a circular economy approach is an opportunity to close the link between food, water, and energy, as well as to incorporate the sustainable perspective to human progress.

Note

Figures and Tables are original from their authorship.

Author details

Alicia Román-Martínez
Universidad Autónoma de San Luis Potosí, Facultad de Ciencias Químicas,
San Luis Potosí, México

*Address all correspondence to: alicia.romanm@uaslp.mx

References

[1] Hamm W, Hamilton RJ, Desmet GC. Edible Oil Processing. 2nd ed. New Jersey: Wiley-Blackwell; 2013

[2] O'Brien RD. Fats and Oils. Formulating and Processing for Applications. 3rd ed. Boca Raton: CRC Press; 2008

[3] USDA. United States Department of Agriculture. 2019. Disponible en: www.fas.usda.gov (Consultado: April 2, 2020)

[4] Nonhebel S. On resource use in food production systems: The value of livestock as "rest-stream upgrading system." Ecological Economics. 2004;**48**(2):221-230

[5] Thomas A, Trentesaux D. Are intelligent manufacturing systems sustainable? Studies in Computational Intelligence. 2014;**544**:3-14

[6] Silk D, Mazzali B, Gargalo CL, Pinelo M, Udugama AI, Mansouri SS. A decision-support framework for techno-economic-sustainability assessment of resource recovery alternatives. Journal of Cleaner Production. 2020;**266**(121854):1-18

[7] Udugama IA, Petersen LAH, Falco FC, Junicke H, Mitic A, Alsina XF, et al. Resource recovery from waste streams in a water-energy-food nexus perspective: Toward more sustainable food processing. Food and Bioproducts Processing. 2020;**119**:133-147

[8] Ghisellini P, Cialani C, Ulgiati S. A review on circular economy: The expected transition to a balanced interplay of environmental and economic systems. Journal of Cleaner Production. 2016;**114**(7):11-32

[9] Maina S, Kachrimanidou V, Koutinas A. A roadmap towards a circular and sustainable bioeconomy through waste valorization. Current Opinion in Green and Sustainable Chemistry. 2017;**8**:18-23

[10] Munir MT, Soheil S, Udugama IA, Baroutian S, Gernaey KV, Young BR. Resource recovery from organic solid waste using hydrothermal processing: Opportunities and challenges. Renewable and Sustainable Energy Reviews. 2018;**96**:64-75

[11] Perederic OA, Mansouri SS, Appel S, Sarup B, Woodley JM, Kontogeorgis GM. Process analysis of Shea butter solvent fractionation using a generic systematic approach. Industrial & Engineering Chemistry Research. 2020;**59**:9152-9164

[12] Sharma YC, Singh B, Madhu D, Liu Y, Yaakob Z. Fast synthesis of high quality biodiesel from "waste fish oil" by single step transesterification. Biofuel Research Journal. 2014;**1**(3):78-80

[13] Venkata Mohan S, Nikhil GN, Chiranjeevi P, Reddy CN, Rohit MV, Kumar AN. Waste biorefinery models towards sustainable circular bioeconomy: Critical review and future perspectives. Bioresource Technology. 2016;**215**:2-12

[14] Jayathilakan K, Sultana K, Radhakrishna K, Bawa AS. Utilization of byproducts and waste materials from meat, poultry and fish processing industries: A review. Journal of Food Science and Technology. 2012;**49**(3):278-293

[15] Eslick GD, Howe PRC, Smith C, Priest R, Bensoussan A. Benefits of fish oil supplementation in hyperlipidemia: A systematic review and meta-analysis. International Journal of Cardiology. 2009;**136**(1):4-16

[16] Bimbo AP. The production and processing of marine oils. In: Breivik H,

editor. Long Chain Omega-3 Specialty Oils. The Oily Press Bridgwater England; 2007. pp. 77-109

[17] Dyerberg J, Bang HO. Haemostatic function and platelet polyunsaturated fatty acids in Eskimos. The Lancet. 1979; **314**(8140):433-435

[18] Farooqui AA. Beneficial Effects of Fish Oil on Human Brain. 1th ed. New York: Springer Science; 2009

[19] Ciriminna R, Meneguzzo F, Delisi R, Pagliaro M. Enhancing and improving the extraction of omega-3 from fish oil. Sustainable Chemistry and Pharmacy. 2017;**5**:54-59

[20] Lands WEM. Fish, Omega-3 and Human Health. 2nd ed. New York: AOCS Press; 2005

[21] Kris-Etherton PM, Harris WS, Appel LJ. Fish consumption, fish oil, omega-3 fatty acids, and cardiovascular disease. Circulation. 2002;**106**(21):2747-2757

[22] Yashodhara BM, Umakanth S, Pappachan JM, Bhat SK, Kamath R, Choo BH. Omega-3 fatty acids: A comprehensive review of their role in health and disease. British Medical Journal. 2009;**85**(March):84-90

[23] Micha R, Khatibzadeh S, Shi P, Fahimi S, Lim S, Andrews KG, et al. Global, regional, and national consumption levels of dietary fats and oils in 1990 and 2010: A systematic analysis including 266 country-specific nutrition surveys. The BMJ. 2014;**348** (g2272):1-20

[24] Grand View Research. Omega 3 Market Size, Share & Trend Analysis Report By Type (EPA, DHA), By Source (Marine, Plant), By Application (Supplements & Functional Foods, Pharmaceuticals, Infant Formulas), By Region, and Segment Forecasts, 2020-2028. (Industry Analysis Report). 2019

[25] Rubio-Rodríguez N, Beltrán S, Jaime I, Diego SM De, Sanz MT, Carballido, JR. Production of Omega-3 polyunsaturated fatty acid concentrates: A review. Innovative Food Science and Emerging Technologies. 2010;**11**(1):1-12

[26] Tena N, Lobo-Prieto A, Aparicio R, García DL. Storage and preservation of fats and oils. In: Encyclopedia of Food Security and Sustainability. Vol. 2. Elsevier Inc.; 2018. pp. 605-618

[27] Damaceno DS, Perederic OA, Ceriani R, Kontogeorgis GM, Gani R. Improvement of predictive tools for vapor-liquid equilibrium based on group contribution methods applied to lipid technology. Fluid Phase Equilibria. 2018; **470**:249-258

[28] Díaz-Tovar CA, Gani R, Sarup B. Lipid technology: Property prediction and process design/analysis in the edible oil and biodiesel industries. Fluid Phase Equilibria. 2011;**302**(1–2):284-293

[29] Perederic OA, Cunico LP, Kalakul S, Sarup B, Woodley JM, Kontogeorgis GM, et al. Systematic identification method for data analysis and phase equilibria modelling for lipids systems. The Journal of Chemical Thermodynamics. 2018;**121**: 153-169

[30] AOCS. Introduction to the Processing of Fats and Oils. Champaign, Illinois: American Oil Chemists' Society; 2003a

[31] Stankiewicz A, Drinkemburg AAH. Process intensification: History, philosophy, principles. In: Stankiewicz A, Moulijn JA, editors. Re-Engineering the Chemical Processing Plant. 2004. pp. 1-32

[32] Freund H, Sundmacher K. Towards a methodology for the systematic analysis and design of efficient chemical processes. Part 1. From unit operations to elementary process functions. Chemical Engineering and Processing: Process Intensification. 2008;**47**(12):2051-2060

[33] Alfa Laval Inc. Multiple Choice for Fats and Oils Refining: Alfa Laval Degumming and Neutralization Solutions. Sweden. 2010 Report; 2010

[34] Svenson E, Willits J. Nano Neutralization. In Green Vegetable Oil Processing. New York: AOCS Press; 2014. pp. 147-157

[35] Ding H, Ye W, Wang Y, Wang X, Li L, Liu D, et al. Process intensification of transesterification for biodiesel production from palm oil: Microwave irradiation on transesterification reaction catalyzed by acidic imidazolium ionic liquids. Energy. 2018;**144**:957-967

[36] Stavarache C, Vinatoru M, Maeda Y. Aspects of ultrasonically assisted transesterification of various vegetable oils with methanol. Ultrasonics Sonochemistry. 2007;**14**(3):380-386

[37] Armenta RE, Vinatoru M, Burja AM, Kralovec JA, Barrow CJ. Transesterification of fish oil to produce fatty acid ethyl esters using ultrasonic energy. Journal of the American Oil Chemists' Society. 2007;**84**(11):1045-1052

[38] Manjula S, Subramanian R. Membrane technology in degumming, dewaxing, deacidifying, and decolorizing edible oils. Critical Reviews in Food Science and Nutrition. 2006;**46**(7):569-592

[39] Ghasemian S, Sahari MA, Barzegar M, Gavlighi HA. Concentration of Omega-3 polyunsaturated fatty acids by polymeric membrane. International Journal of Food Science and Technology. 2015;**50**(11):2411-2418

[40] Ivanovs K, Blumberga D. Extraction of fish oil using green extraction methods: A short review. Energy Procedia. 2017;**128**:477-483

[41] Ferdosh S, Sarker ZI, Norulaini N, Oliveira A, Yunus K, Chowdury AJ, et al. Quality of tuna fish oils extracted from processing the by-products of three species of neritic tuna using supercritical carbon dioxide. Journal of Food Processing and Preservation. 2015;**39**(4):432-441

[42] Fiori L, Volpe M, Lucian M, Anesi A, Manfrini M, Guella G. From fish waste to Omega-3 concentrates in a biorefinery concept. Waste and Biomass Valorization. 2017;**8**(8):2609-2620

[43] Riha V, Brunner G. Separation of fish oil ethyl esters with supercritical carbon dioxide. Journal of Supercritical Fluids. 2000;**17**(1):55-64

[44] Oterhals Å, Kvamme B, Berntssen MHG. Modeling of a short-path distillation process to remove persistent organic pollutants in fish oil based on process parameters and quantitative structure properties relationships. Chemosphere. 2010;**80**(2):83-92

[45] Rossi P, Grosso NR, Pramparo M, Del C, Nepote V. Fractionation and concentration of Omega-3 by molecular distillation. In: Bradley TG, Vargas FP, editors. Eicosapentaenoic Acid: Sources, Health Effects and Role in Disease Prevention. 1st ed. Billerica, MA: Nova Biomedical; 2012. pp. 177-203

[46] Solaesa ÁG, Sanz MT, Falkeborg M, Beltrán S, Guo Z. Production and concentration of monoacylglycerols rich in omega-3 polyunsaturated fatty acids by enzymatic glycerolysis and molecular distillation. Food Chemistry. 2016;**190**:960-967

[47] Bernal JL, Martín MT, Toribio L. Supercritical fluid chromatography in food analysis. Journal of Chromatography A. 2013;**1313**:24-36

[48] Dołowy M, Pyka A. Chromatographic methods in the separation of long-chain mono- and polyunsaturated fatty acids. Journal of Chemistry. 2015;**2015**:1-20

[49] Fagan P, Wijesundera C. Rapid isolation of Omega-3 long-chain polyunsaturated fatty acids using monolithic high performance liquid chromatography columns. Journal of Separation Science. 2013;**36**(11): 1743-1752

[50] Grace & Co., W. R. TRISYL® Silica Gel for Refining Edible Oil. 2011. Disponible en: www.grace.com/food-and-beverage/en-US/edible-oil-refining-aids

[51] Feng X, Patterson DA, Balaban M, Fauconnier G, Emanuelsson EAC. The spinning cloth disc reactor for immobilized enzymes: A new process intensification technology for enzymatic reactions. Chemical Engineering Journal. 2013;**221**:407-417

[52] Xu X. Engineering of enzymatic reactions and reactors for lipid modification and synthesis. European Journal of Lipid Science and Technology. 2003;**105**(6):289-304

[53] Yan X, Zhao X, Ma G, Dai L, Du W, Liu D. Enzymatic ethanolysis of fish oil for selective concentration of polyunsaturated fatty acids (PUFAs) with flexible production of corresponding glycerides and ethyl esters. Journal of Chemical Technology and Biotechnology. 2018;**93**(8): 2399-2405

[54] Stavarache C, Vinatoru M, Nishimura R, Maeda Y. Fatty acids methyl esters from vegetable oil by means of ultrasonic energy. Ultrasonics Sonochemistry. 2005;**12**(5):367-372

[55] Linder M, Matouba E, Fanni J, Parmentier M. Enrichment of salmon oil with n-3 PUFA by lipolysis, filtration and enzymatic re-esterification. European Journal of Lipid Science and Technology. 2002;**104**(8):455-462

[56] Carvalho A, Mimoso AF, Mendes AN, Matos HA. From a literature review to a framework for environmental process impact assessment index. Journal of Cleaner Production. 2014;**64**:36-62

[57] Lewandowski M. Designing the business models for circular economy-towards the conceptual framework. Sustainability. 2016;**8**(1):1-28

[58] Khatri P, Jain S. Environmental life cycle assessment of edible oils: A review of current knowledge and future research challenges. Journal of Cleaner Production. 2017;**152**:63-76

[59] Lutze P, Román-Martinez A, Woodley JM, Gani R. A systematic synthesis and design methodology to achieve process intensification in (bio) chemical processes. Computers and Chemical Engineering. 2012;**36**(1):189-207

[60] Mota B, Gomes MI, Carvalho A, Barbosa-Povoa AP. Towards supply chain sustainability: Economic, environmental and social design and planning. Journal of Cleaner Production. 2015;**105**:14-27

[61] Yee KF, Tan KT, Abdullah AZ, Lee KT. Life cycle assessment of palm biodiesel: Revealing facts and benefits for sustainability. Applied Energy. 2009; **86**:S189-S196

[62] Othman MR, Repke JU, Wozny G, Huang YL. A modular approach to

sustainability assessment and decision support in chemical process design. Industrial & Engineering Chemistry Research. 2010;**49**(17):7870-7881

[63] Landucci G, Nucci B, Pelagagge L, Nicolella C. Hazard assessment of edible oil refining: Formation of flammable mixtures in storage tanks. Journal of Food Engineering. 2011;**105**: 105-111

[64] Morais S, Mata TM, Martins AA, Pinto GA, Costa CAV. Simulation and life cycle assessment of process design alternatives for biodiesel production from waste vegetable oils. Journal of Cleaner Production. 2010;**18**(13):1251-1259

[65] Lee S, Posarac D, Ellis N. Process simulation and economic analysis of biodiesel production processes using fresh and waste vegetable oil and supercritical methanol. Chemical Engineering Research and Design. 2011; **89**(12):2626-2642

[66] Tran NN, Tišma M, Budžaki S, McMurchie EJ, Gonzalez OMM, Hessel V, et al. Scale-up and economic analysis of biodiesel production from recycled grease trap waste. Applied Energy. 2018; **229**:142-150

[67] Schmidt JH. Life cycle assessment of five vegetable oils. Journal of Cleaner Production. 2015;**87**(C):130-138

[68] Banchero M, Kusumaningtyas RD, Gozzelino G. Reactive distillation in the intensification of oleic acid esterification with methanol - a simulation case-study. Journal of Industrial and Engineering Chemistry. 2014;**20**(6):4242-4249

[69] Likozar B, Pohar A, Levec J. Transesterification of oil to biodiesel in a continuous tubular reactor with static mixers: Modelling reaction kinetics, mass transfer, scale-up and optimization considering fatty acid composition. Fuel Processing Technology. 2016;**142**: 326-336

[70] Martín M, Grossmann IE. Simultaneous optimization and heat integration for the coproduction of diesel substitutes: Biodiesel (FAME and FAEE) and glycerol ethers from algae oil. Industrial and Engineering Chemistry Research. 2014;**53**(28):11371-11383

[71] Diaz-Tovar CA, Ceriani R, Gani R, Bent S. Systematic methodology and property prediction of fatty Systems for Process Design/analysis in the oil and fat industry. Brazilian Journal of Chemical Engineering. 2010;**27**(03):401-412

[72] Shabbir Z, Tay DHS, Ng DKS. A hybrid optimisation model for the synthesis of sustainable gasification-based integrated biorefinery. Chemical Engineering Research and Design. 2012; **90**(10):1568-1581

[73] Jones M, Forero-Hernandez H, Zubov A, Sarup B, Sin G. Superstructure optimization of Oleochemical processes with surrogate models. Computer Aided Chemical Engineering. 2018;**44**(1):277-282

[74] Kamarden H, Kidam K, Hashim H, Hassim MH, Shahlan SS, Ngadi N, et al. Process simulation of integrated palm oil mill, refinery and oleochemical processes. IOP Conf. Series. Materials Science and Engineering. 2018;**458** (012062):1-6

[75] Barr WJ, Landis AE. Comparative life cycle assessment of a commercial algal multiproduct biorefinery and wild caught fishery for small pelagic fish. International Journal of Life Cycle Assessment. 2018;**23**(5):1141-1150

[76] Sotoft LF, Rong B, Christensen KV, Norddahl B. Process simulation and

economical evaluation of enzymatic biodiesel production plant. Bioresource Technology. 2010;**101**(14):5266-5274

[77] Akoh CC, Min DB. Food Lipids: Chemistry, Nutrition and Biotechnology. 2th ed. New York: Marcel Dekker Inc.; 2002

[78] Gupta MK. Practical Guide to Vegetable Oil Processing. 2nd ed. Elsevier Inc: AOCS Press; 2017

[79] Lee SJ, Ying DY. Encapsulation of fish oils. In: Garti N, editor. Delivery and Controlled Release of Bioactives in Foods and Nutraceuticals. Sawston, Cambridge: Woodhead Publishing Limited; 2008. pp. 370-403

[80] Bhosle BM, Subramanian R. New approaches in deacidification of edible oils - a review. Journal of Food Engineering. 2005;**69**(4):481-494

[81] Bimbo AP. Guidelines for characterizing food-grade fish oil. Inform. 1998;**9**(5):473-483

[82] GOED. GOED best practice guidelines: oxidation control (Reporte 2017). 2017

[83] Blanco-Rodríguez P. Diseño de una planta piloto de refinación de aceites vegetales. (tesis de pregrado). España: Universidad de Cádiz; 2007

[84] Bonilla-Mendez JR, Hoyos-Concha JL. Methods of extraction, refining and concentration of fish oil as a source of omega-3 fatty acids. Corpoica Ciencia y Tecnologia Agropecuaria. 2018;**19**(3): 645-668

[85] Mensink RP, Zock PL, Kester ADM, Katan MB. Effects of dietary fatty acids and carbohydrates on the ratio of serum total to HDL cholesterol and on serum lipids and apolipoproteins: A meta-analysis of 60 controlled trials. American Journal of Clinical Nutrition. 2003;**77**(5):1146-1155

[86] Yves H, Korma SA, Ali AH, Tuyishime MA, Habinshuti I, Abed SM. Extraction, refining and purification of ω-3 PUFA through different Techniques – A review. American Journal of Food Science and Nutrition Research. 2016;**4**(1):18-26

[87] International Standards Organization. ISO 5555:2001. Animal and vegetable fats and oils - Sampling. 2001

[88] AOCS. Official method Ca 5a-40. Sampling and analysis of commercial fats and oils: Free fatty acids. American Oil Chemists' Society. 2009a

[89] AOCS. Official method Cd 18-90. Sampling and analysis of commercial fats and oils: Anisidine value. American Oil Chemists' Society. 2003b

[90] AOCS. Official method Cd 8b-90. Sampling and analysis of commercial fats and oils: Peroxide value. American Oil Chemists' Society. 2003d

[91] AOCS. Official method Cd 1d-92. Sampling and analysis of commercial fats and oils: Lodine value of fats and oils cyclohexane-acetic acid method. American Oil Chemists' Society. 2003c

[92] AOCS. Official method Ce 1b-89. Sampling and analysis of commercial fats and oils: Fatty acid composition of marine oils by GLC. American Oil Chemists' Society. 2009b

[93] Young FVK. The Chemical & Physical Properties of crude fish oils for refiners & Hydrogenators. Fish oil bulletin. International Fishmeal and Fishoil Organization (IFFO). 1986;**18**: 1-18

[94] Suseno SH, Sri Hayati S, Izaki AF. Fatty acid composition of some potential fish oil from production centers in Indonesia. Oriental Journal of Chemistry. 2014;**30**(3):975-980

[95] FAO. Standard for Fish Oils. Food and Agriculture Organization of the United Nations. Codex Alimentarius Commission Report STAN 329-2017. 2017

[96] Schuchardt JP, Hahn A. Bioavailability of long-chain omega-3 fatty acids. Prostaglandins Leukotrienes and Essential Fatty Acids. 2013;**89**(1):1-8

[97] Zhang H, Shen Y, Zhang Y, Li L, Wang X. Regiospecific analysis of fatty acids and calculation of triglyceride molecular species in marine fish oils. BioMed Research International. 2018; **2018**:1-7

[98] Carvajal AK, Mozuraityte R. Fish oils: Production and properties. Encyclopedia of Food and Health. 2016: 693-698

[99] Oterhals Å, Vogt G. Impact of extraction, refining and concentration stages on the stability of fish oil. In: Food Enrichment with Omega-3 Fatty Acids. Sawston, Cambridge: Woodhead Publishing; 2013. pp. 111-129

[100] Ismail A, Bannenberg G, Rice HB, Schutt E, Mackay D. Oxidation in EPA and DHA rich oils: An overview. Lipid Technology. 2016;**28**(3–4):55-59

[101] Laguerre M, Lecomte J, Villeneuve P. Evaluation of the ability of antioxidants to counteract lipid oxidation: Existing methods, new trends and challenges. Progress in Lipid Research. 2007;**46**(5):244-282

[102] Omegavia. How to Buy the Best Fish Oil Supplements – Part 3: Freshness. 2022. Disponible en https://omegavia.com/best-fish-oil-supplement-3/

[103] Cunico LP, Ceriani R, Sarup B, O'Connell JP, Gani R. Data, analysis and modeling of physical properties for process design of systems involving lipids. Fluid Phase Equilibria. 2014;**362**: 318-327

[104] Cunico LP, Hukkerikar AS, Ceriani R, Sarup B, Gani R. Molecular structure-based methods of property prediction in application to lipids: A review and refinement. Fluid Phase Equilibria. 2013; **357**:2-18

[105] Diaz-Tovar CA, Mustaffa AA, Hukkerikar A, Quaglia A, Sin G, Kontogeorgis G, et al. Lipid processing technology: Building a multilevel modeling network. Computer Aided Chemical Engineering. 2011;**29**: 256-260

[106] Pereira E, Meirelles AJA, Maximo GJ. Predictive models for physical properties of fats, oils, and biodiesel fuels. Fluid Phase Equilibria. 2020;**508** (112440):1-20

[107] Lydersen AL. Estimation of critical properties of organic compounds. In: College Engineering University Wisconsin, Engineering Experimental Station, Report 3. Madison, WI: April; 1955

[108] Ambrose D. The correlation and estimation of vapor–liquid critical properties. I. Critical temperatures of organic components. The Journal of Chemical Thermodynamics. Teddington, UK. 1978;**10**(8):765-769

[109] Klincewicz KM, Reid RC. Estimation of critical properties with group contribution methods. AIChE Journal. 1984;**30**(1):137-142

[110] Chein-Hsiun T. Group-contribution estimation of critical temperature with only chemical structure. Chemical Engineering Science. 1995;**50**(22): 3515-3520

[111] Joback KG, Reid RC. Estimation of pure component properties from group contributions. Chemical Engineering Communications. 1987;**57**(1–6):233-243

[112] Constantinou L, Gani R. New group contribution method for estimating properties of pure compounds. AICHE Journal. 1994;**40**(10):1697-1710

[113] Gani R, Marrero J. Group-contribution based estimation of pure component properties. Fluid Phase Equilibria. 2001;**183**:183-208

[114] Kolská Z, Ruzicka V, Gani R. Estimation of the enthalpy of vaporization and the entropy of vaporization for pure organic compounds at 298.15 K and at Normal boiling temperature by a group contribution method. Industrial & Engineering Chemical Research. 2005; **44**:8436-8454

[115] Kolska Z, Kukal J, Zábranský M, Ruzicka V. Estimation of the heat capacity of organic liquids as a function of temperature by a three-level group contribution method. Industrial & Engineering Chemical Research. 2008; **47**:2075-2085

[116] Ceriani R, Gani R, Meirelles AJA. Prediction of heat capacities and heats of vaporization of organic liquids by group contribution methods. Fluid Phase Equilibria. 2009;**283**(1–2):49-55

[117] Ceriani R, Gonçalves CB, Coutinho JAP. Prediction of viscosities of fatty compounds and biodiesel by group contribution. Energy and Fuels. 2011; **25**(8):3712-3717

[118] Ceriani R, Gani R, Liu YA. Prediction of vapor pressure and heats of vaporization of edible oil/fat compounds by group contribution. Fluid Phase Equilibria. 2013;**337**:53-59

[119] Zeberg-Mikkelsen CK, Stenby EH. Predicting the melting points and the enthalpies of fusion of saturated triglycerides by a group contribution method. Fluid Phase Equilibria. 1999; **162**:7-17

[120] Moorthy AS, Liu R, Mazzanti G, Wesdorp LH. Estimating thermodynamic properties of pure triglyceride systems using the triglyceride property calculator. Journal of the American Oil Chemists' Society. 2016;**94**(2):187-199

[121] Wittig R, Lohmann J, Gmehling J. Vapor-liquid equilibria by UNIFAC group contribution. 6. Revision and extension. Industrial and Engineering Chemistry Research. 2003;**42**(1): 183-188

[122] Magnussen T, Rasmussen P, Fredenslund A. Unifac parameter table for prediction of liquid-liquid equilibria. Industrial and Engineering Chemistry Process Design and Development. 1981; **20**(2):331-339

[123] Gmehling J, Li J, Schiller M. A modified UNIFAC model. 2. Present parameter matrix and results for different thermodynamic properties. Industrial and Engineering Chemistry Research. 1993;**32**(1):178-193

[124] Hukkerikar AS, Sarup B, Ten Kate A, Abildskov J, Sin G, Gani R. Group-contribution + (GC +) based estimation of properties of pure components: Improved property estimation and uncertainty analysis. Fluid Phase Equilibria. 2012;**321**:25-43

Chapter 2

Biotechnology for Improving Hydroxy Fatty Acids Production in Lesquerella (*Physaria fendleri*)

Grace Chen and Kumiko Johnson

Abstract

Hydroxy fatty acid (HFA) is a vital raw material for numerous industrial products, such as lubricants, plasticizers and surfactants. Castor oil is the current commercial source of HFA which contains 90% ricinoleic acid (18,1OH). Castor seeds contain the toxin ricin and hyperallergic 2S albumins; it is detrimental to castor oil production. Lesquerella is a potential industrial oilseed crop for a safe source of HFA, because lesquerella seeds contain a valuable HFA, lesquerolic acid (20,1OH), at 55–60% in seed oil. This chapter describes current progress on improving HFA production in lesquerella through metabolic engineering.

Keywords: hydroxy fatty acid, ricinoleic acid, lesquerolic acid, triacylglycerol, *Physaria fendleri*, lesquerella, seed oil, genetic transformation

1. Introduction

Lesquerella seed oil (triacylglycerol, TAG) biosynthesis follows common *de novo* fatty acid (FA) biosynthesis in plastid (**Figure 1**). Once oleic acid (18,1) is synthesized and exported to cytosol, it is activated to 18:1-Coenzyme A (CoA) for endoplasmic reticulum (ER)-mediated fatty acid modification and TAG assembly [1]. The 18:1-CoA can be esterified directly into membrane lipid phosphatidylcholine (PC) in the ER by the forwarding reaction of lyso-PC acyltransferase (LPCAT) [2–4] resulting in 18:1-PC (**Figure 1**). An oleate 12-hydroxylase (FAH12) [5–8] hydroxylates 18:1-PC to form 18:1OH-PC (**Figure 1**). Lesquerella PfFAH12, however, converts 18:1-PC to both 18:1OH-PC and linoleic acid (18,2)-PC, because PfFAH12 posesses bi-functional FAD2-related oleate Δ12 - hydroxylase: desaturase activities [8].

Through the reverse reaction of LPCAT (**Figure 1**), or phospholipase A (PLA2)–type activity [9], the 18:1OH can be removed from PC and transferred back to cytosol to be activated as 18:1OH-CoA. A lesquerella fatty acid condensing enzyme (PfKCS18) (also called KCS3 or FAE1) elongates 18:1OH-CoA to 20:1OH-CoA [10] (**Figure 1**). Rapid acylation of 18:1 and de-acylation of 18:1OH by LPCAT (or by PLA2), and together with elongation of 18:1OH-CoA by PfKCS18 leads to enrichment of 20:1OH-CoA in cytosol. FA desaturase 2 (FAD2) [11] and FA desaturase 3 (FAD3) [12] converts 18:1-PC to 18:2-PC and 18:2-PC to linolenic acids (18:3)-PC,

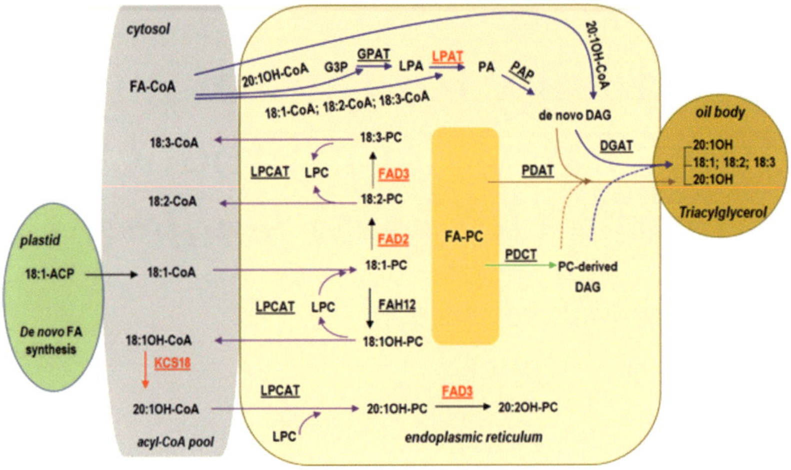

Figure 1.
Proposed pathways for TAG biosynthesis in lesquerella seed. Kennedy pathway is indicated by blue arrows. Acyl editing reactions are indicated by purple arrows. PC-derived DAG formation is indicated by a green arrow. Enzymes catalyzing these reactions are underlined. Fatty acid numerical symbols and abbreviations are described in introduction.

respectively (**Figure 1**). The FAD3 is also resposible for converting 20-1OH to auricolic acid (20,2OH) [13, 14] in lesquerella seeds. A functional lesquerella PfFAD3–1 isoform is a key enzyme producing 18:3 and 20:2OH [15]. FA-CoA or FA-PC are assembled to TAG through multiple mechanisms [1, 2]. First, FA-CoAs are acylated to a glycerol-3-phosphate (G3P) backbone by enzymes in Kennedy pathway [16]. Glycerol-3-phosphate acyltransferase (GPAT) acylates FA-CoA to the *sn*-1 position of G3P backbone forming lysophosphatidic acid (LPA). LPA acyltransferase (LPAT) is responsible to add another FA-CoA to the *sn*-2 of LPA forming phosphatidic acid (PA). Through PA phosphatase (PAP), PA is changed to 1,2-*sn*-diacylglycerol (DAG). The DAG produced in Kennedy pathway is referred to de novo DAG. The last FA-CoA can be added to the *sn*-3 of DAG by 1,2-*sn*-diacylglycerol acyltransferase (DGAT), creating TAG. In lesquerella, almost all seed TAG having 20:1OH at the *sn*-1 and *sn*-3 positions, and the *sn*-2 positions are occupied by unsaturated FAs, ie., 18:1, 18:2 and 18:3 [17–20]. The reason of lack of HFA at the *sn*-2 position of TAG could be due to the selectivity of lesquerella LPAT (PfLPAT2) for unsaturated FA [21], which is a typical characteristic for most plant LPAT2 [22]. Second, PC can be converted to DAG (PC-derived DAG). PC:DAG cholinephosphotransferase (PDCT) [23–25] is a major enzyme to produce PC-derived DGA through exchange of the head group between PC and DAG. Alternatively, PC-derived DGA can be produced by other enzymatic reactions catalyzed by CDP-choline: DAG cholinephosphotransferase (CPT) [26], or phospholipases (PLC, or PLD) [2, 27]. The conversion of PC into DAG also provides a mechanism to increase the amount of 18:1, 18:2, 18:3 in *sn*-2-TAG. Third, FA on the *sn*-2 PC can be transferred to the *sn*-3 position of DAG by phospholipid:DAG acyltransferase (PDAT) (**Figure 1**) [28–30].

To develop lesquerella that produces 18:1OH-rich seed oils like castor, we have over-expressed castor RcLPAT2 [21]. The resulted transgenic lesquerella seeds increase 18:1OH content at the *sn*-2 position of TAG from 2–17%, and consequently, oil accumulates more TAGs with all three *sn* positions occupied by HFA [20, 21]. RNA interference sequences targeting KCS18, FAD2 and FAD3 have been introduced to lesquerella. Seeds from the transgenic lines had increased 18:1OH up to 26.6% compared with that of 0.4–0.6% in wild type (WT) seeds. Our studies enhance our understanding of plant lipid metabolism.

2. Castor LPAT2 increases castor oil-like triacylglycerols in lesquerella seed

We have produced 17 transgenic lesquerella lines expressing *RcLPAT2* under the control of a seed specific promoter [21]. Our results indicate that RcLPAT2 enables the incorporation of 18:1OH at *sn*-2 position of LPA which increases the accumulation of 18:1OH and also tri-HFA-TAGs in lesquerella (**Figures 2** and **3**).

In transgenic lesquerella expressing *RcLPAT2*, we observed an increase in 0-, 1-, and 3-HFA-TAG levels and a reduction in 2-HFA-TAG species (**Figure 2**). Regiochemical analysis showed that *sn*-2 18:1OH increased 6–7-fold or 1.5-fold, respectively (**Figure 3**). Further analysis of regioisomer of the transgenic seed oil reveals that RcLPAT2 increased 3-HFA-TAG content by acylating mostly 18:1OH at the *sn*-2 position of 20:1OH-LPA forming tri-HFA-TAG (20,1OH at *sn*-1, 3 and 18:1OH at *sn*-2) [18–20, 31]. This indicates that *RcLPAT2* allows for a more efficient acylation of 18:1OH than 20:1OH to the *sn*-2 position of TAG *in vivo*. We have demonstrated that castor LPAT2 increases 18:1OH level through exclusively acylating 18:1OH at the *sn*-2 position of tri-HFAs-TAGs in lesquerella. RcLPAT2 holds a valuable property for the engineering of a new castor oil-producing crop, such as lesquerella.

3. Ricinoleic acid content can be increased in lesquerella seeds through suppressing an elongase and fatty acid desaturases

To develop lesquerella that produces 18:1OH-rich seed oils like castor, silencing PfKCS18 would result in accumulation of 18:1OH; silencing FAD2 and FAD3 would block 18:1 flux to 18:2 and 18:3, respectively. To test the hypothesis, we generated transgenic lines expressing RNAi of camelina CsFAD2, CsFAD3, and Arabidopsis

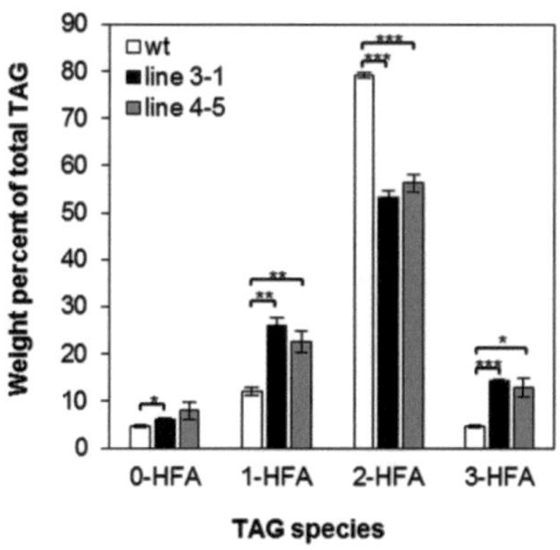

Figure 2.
TAG species composition. Triplicates of 50-seed sample were measured for wild type (wt) and transgenic lines (line 3–1, line 4–5). The data represent averages of three replicates ± SE. Two-tailed Student's t test. $^{*}p < 0.05$; $^{**}p < 0.01$; $^{***}p < 0.001$. *0-HFA, 1-HFA, 2-HFA, and 3-HFA indicate TAG molecular with zero, one, two, or three HFAs, respectively.*

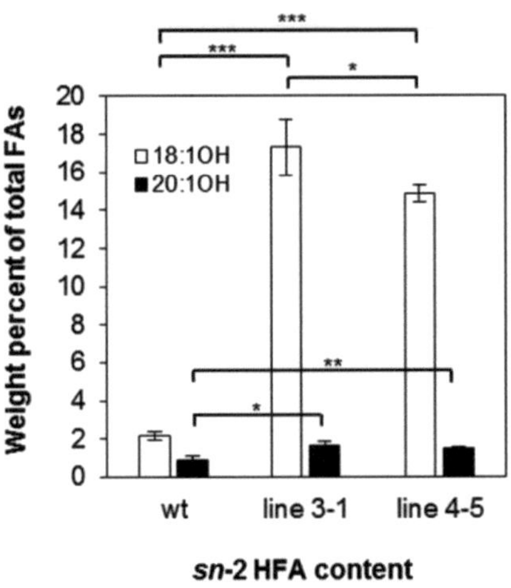

Figure 3.
HFA content at the sn-2 position of TAG. Triplicates of 50-seed sample were measured for wild type (wt) and transgenic lines (line 3–1, line 4–5). The data represent averages of three replicates ± SE. Two-tailed Student's t test. $^{*}p < 0.05$; $^{**}p < 0.01$; $^{***}p < 0.001$. *18:1OH and 20:1OH represent ricinoleic acid and lesquerolic acid, respectively.*

AtKCS18, which have 82–95% sequence homology with corresponding lesquerella genes [18]. RNAi constructs, *CsFAD2 RNAi*, *CsFAD3 RNAi* and *AtFEA1 RNAi*, are effective in silencing corresponding gene expression in camelina [32–34]. We therefore generated 16 transgenic lesquerella lines expressing *AtFAD3 RNAi* + *CsFAE1 RNAi* (2-dsRNA) (**Table 1**) [35], and 15 lines expressing *CsFAD2 RNAi* + *AtFAD3 RNAi* + *CsFAE1 RNAi* (**Table 2**) [35].

As shown in **Table 1**, when the 2-dsRNAs (*AtFAD3 RNAi* and *CsFAE1 RNAi*) was introduced to lesquerella, we observed significant increases in 18:1OH from 0.6% of WT to 26.6% in line 1 (**Table 1**) and decreases in 20:1OH from 51.2% of WT to lowest 19% in line 1 (**Table 1**). 18:1 content was increased in 50% of transgenic population with the highest level of 32.1% in line 2 compared with 17% of WT (**Table 1**). Correlation analysis was performed to show the relationships between FA accumulation for 2-dsRNA group. Among the 16 T_1 transgenic lesquerella lines, 15 lines shifted the accumulation of 18:3 to 18:2, showing a strong negative correlation between 18:2 and 18:3 ($r = 0.93$); 13 lines shifted 20:1OH to 18:1OH, which also displayed a strong negative correlation ($r = -0.99$). These results indicate that *AtFAD3 RNAi* and *CsFAE1 RNAi* are effective in silencing *PfFDA3–1* and *PfKCS18*, respectively.

As shown in **Table 2**, 15 independent transgenic lines expressing the 3-dsRNAs, *CsFAD2 RNAi* + *AtFAD3 RNAi* + *CsFAE1 RNAi* were generated and their T_1 seeds were analyzed for FA composition. Compared with 2-dsRNAs (**Table 1**), similar average contents in 18:2, 18:3 and 20:1 in lines expressing 3-dsRNA (**Table 2**) were observed. With the addition of *CsFAD2 RNAi* in the 3-dsRNA group, the average of 18:1 was higher at 27.8% (**Table 2**) compared with the average of 20.5% in the 2-dsRNA group (**Table 1**). Besides, less dynamic changes between average increase of 18:1OH and average decrease of total HFA were observed in the 3-dsRNA group, showing averages of 4.7% and 48.9%, respectively, (**Table 2**), compared with that of 7.7% and 53% in lines expressing 2-dsRNA, respectively (**Table 1**). FA composition in WT seeds (**Tables 1** and **2**) were similar to described [21, 36]. There was no change of growth phenotype for transgenic lesquerella expressing *CsFAD2 RNAi*, *CsFAD3 RNAi* and *AtFEA1 RNAi*.

Line	Total minor fatty acid[a]	18:1	18:2	18:3	18:1OH	20:1OH	20:2OH	Total hydroxy fatty acid
wild-type	4.7 ± 0.4	17.0 ± 0.4	7.6 ± 0.4	13.3 ± 0.6	0.6 ± 0.2	51.2 ± 1.0	4.3 ± 0.6	56.0 ± 0.5
line 1	4.3 ± 0.2	30.5 ± 2.8 ***	16.2 ± 1.7 ***	2.2 ± 0.7 ***	26.6 ± 0.2 ***	19.0 ± 2.0 ***	0.2 ± 0.2 ***	45.8 ± 1.9 ***
line 2	5.1 ± 0.2	32.1 ± 1.3 **	17.0 ± 0.5 ***	1.7 ± 0.5 ***	16.8 ± 0.5 ***	26.1 ± 1.0 ***	0.0 ± 0.0 ***	42.9 ± 0.5 ***
line 3	5.0 ± 0.2	25.5 ± 1.9 **	17.6 ± 1.2 ***	2.6 ± 1.4 ***	16.6 ± 1.0 ***	31.4 ± 2.1 ***	0.5 ± 0.4 ***	48.4 ± 1.5 ***
line 4	4.7 ± 0.4	22.7 ± 2.7 *	16.8 ± 0.6 ***	3.3 ± 0.8 ***	11.7 ± 2.7 **	39.3 ± 5.0 **	0.4 ± 0.2 ***	51.4 ± 2.6 *
line 5	4.7 ± 0.2	23.4 ± 2.0 **	18.7 ± 0.2 ***	1.8 ± 0.6 ***	10.2 ± 0.4 ***	40.2 ± 1.4 ***	0.1 ± 0.1 ***	50.5 ± 1.5 ***
line 6	4.4 ± 0.1	20.5 ± 1.8 *	14.4 ± 1.0 ***	6.2 ± 0.5 ***	8.5 ± 0.4 ***	43.7 ± 1.4 **	1.2 ± 0.1 ***	53.4 ± 1.5 *
line 7	4.1 ± 0.1	20.5 ± 0.9 **	16.7 ± 2.0 ***	4.9 ± 1.7 ***	8.0 ± 1.3 ***	45.5 ± 2.0 **	0.3 ± 0.2 ***	53.8 ± 1.2 *
line 8	4.5 ± 0.0	17.6 ± 0.5	16.1 ± 0.5 ***	7.6 ± 0.4 ***	7.5 ± 0.5 ***	45.0 ± 0.5 ***	1.3 ± 0.2 ***	53.8 ± 0.2 **
line 9	4.4 ± 0.1	17.9 ± 0.5	17.2 ± 0.7 ***	4.3 ± 0.7 ***	4.9 ± 0.4 ***	49.3 ± 0.6 *	0.7 ± 0.2 ***	54.9 ± 0.5
line 10	4.8 ± 0.0	19.0 ± 1.0 *	13.0 ± 0.5 ***	9.6 ± 0.5 ***	4.7 ± 1.2 **	46.9 ± 0.6 *	1.2 ± 0.1 ***	52.8 ± 1.4 *
line 11	4.3 ± 0.0	16.5 ± 0.4	16.3 ± 0.2 ***	5.4 ± 0.3 ***	3.6 ± 0.3 ***	51.9 ± 0.4	0.8 ± 0.2 ***	56.2 ± 0.0
line 12	4.1 ± 0.2	16.3 ± 0.5	14.1 ± 0.9 ***	7.7 ± 1.2 **	1.8 ± 0.3 **	53.2 ± 1.0	1.8 ± 0.2 **	56.7 ± 1.1
line 13	3.9 ± 0.3	17.1 ± 2.1	9.5 ± 0.2 **	13.3 ± 0.4	1.2 ± 0.1 **	51.1 ± 2.7	2.2 ± 0.1 **	54.5 ± 2.7
line 14	4.3 ± 0.3	16.2 ± 0.5	20.0 ± 0.7 ***	1.5 ± 0.1 ***	0.5 ± 0.1	56.2 ± 1.2 **	0.1 ± 0.0 ***	56.8 ± 1.3
line 15	4.0 ± 0.4	16.1 ± 0.5	14.2 ± 1.2 ***	6.6 ± 0.9 ***	0.5 ± 0.1	55.5 ± 0.7 **	1.8 ± 0.4 **	57.7 ± 0.8 *
line 16	4.1 ± 0.1	15.4 ± 0.4 **	13.4 ± 1.5 **	8.3 ± 1.3 **	0.5 ± 0.1	55.6 ± 0.2 **	1.7 ± 0.3 **	57.8 ± 0.5 *
average of transgenics	4.4 ± 0.1	20.5 ± 0.9	15.7 ± 0.5	5.4 ± 0.4	7.7 ± 0.7	44.4 ± 1.2	0.9 ± 0.1	53.0 ± 0.8

Three or four replicates of 30-seed sample were measured for wild-type and each transgenic line. All data are averages of measurements ±SD. Fatty acid legend: 18:1 is oleic; 18:2 is linoleic; 18:3 is linolenic; 18:1OH is ricinoleic; 20:1OH is lesquerolic; and 20:2OH is auricolic acid. a, total content of five common fatty acids: palmitic (16:0), palmitoleic (16:1), stearic (18:0), arachidic (20:0), and eicosenoic acids (20:1). Two-tailed Student's t-test. $^{*}p < 0.05$.
$^{**}p < 0.01$.
$^{***}p < 0.001$.

Table 1.
Fatty acid composition (mole %) in T_1 *seeds expressing AtFAD3 RNAi + CsFAE1 RNAi.*

Line	Total minor fatty acid[a]	18:1	18:2	18:3	18:1OH	20:1OH	20:2OH	Total hydroxy fatty acid
wild-type	4.2 ± 0.2	16.7 ± 0.2	8.0 ± 0.2	13.7 ± 0.2	0.40 ± 0.0	53.0 ± 0.9	3.1 ± 0.5	56.5 ± 0.7
line 1	4.2 ± 0.1	27.7 ± 0.3 ***	13.8 ± 0.4 ***	4.8 ± 0.5 ***	15.4 ± 0.7 ***	33.3 ± 0.8 ***	0.9 ± 0.1 ***	49.6 ± 0.1 ***
line 2	4.3 ± 0.1	26.4 ± 2.4 **	13.6 ± 0.3 ***	5.5 ± 0.8 ***	10.3 ± 0.9 ***	38.8 ± 2.5 ***	1.1 ± 0.1 **	50.1 ± 1.9 **
line 3	5.2 ± 0.0 ***	35.7 ± 1.6 **	15.1 ± 0.4 ***	3.1 ± 1.0 ***	8.2 ± 1.1 ***	32.4 ± 1.8 ***	0.5 ± 0.3 **	40.9 ± 1.1 ***
line 4	4.1 ± 0.3	22.6 ± 0.6 ***	15.8 ± 1.5 ***	4.3 ± 2.5 **	7.5 ± 1.0 ***	44.9 ± 0.7 ***	0.7 ± 0.6 **	53.2 ± 1.3 **
line 5	5.4 ± 0.1 ***	30.2 ± 1.0 ***	13.2 ± 1.1 ***	3.8 ± 0.3 ***	6.5 ± 0.8 ***	37.3 ± 1.0 ***	0.7 ± 0.1 ***	44.5 ± 1.8 ***
line 6	4.5 ± 0.0 *	35.8 ± 4.1 ***	15.1 ± 0.2 ***	5.0 ± 0.7 ***	6.3 ± 0.4 ***	35.6 ± 3.7 ***	0.7 ± 0.3 ***	42.6 ± 4.1 **
line 7	4.1 ± 0.1	17.5 ± 0.8	17.3 ± 1.2 ***	3.9 ± 0.9 ***	6.3 ± 0.5 ***	50.2 ± 0.5 **	0.7 ± 0.0 ***	57.2 ± 0.6
line 8	4.3 ± 0.2 *	38.8 ± 3.5 ***	14.4 ± 0.9 ***	1.7 ± 0.3 ***	4.6 ± 0.3 ***	36.1 ± 2.8 ***	0 ± 0.3 ***	40.7 ± 2.6 ***
line 9	3.9 ± 0.2	24.6 ± 2.1 **	13.3 ± 0.5 ***	6.0 ± 0.5 ***	1.9 ± 0.2 ***	48.9 ± 1.4 **	1.2 ± 0.2 **	52.1 ± 1.4 **
line 10	4.5 ± 0.8	32.3 ± 2.8 ***	10.7 ± 1.3 ***	7.1 ± 1.1 ***	1.1 ± 0.2 ***	43.1 ± 2.2 **	1.5 ± 0.3 **	45.7 ± 1.8 ***
line 11	4.8 ± 0.3 *	22.9 ± 0.8 ***	13.9 ± 0.1 ***	7.0 ± 0.7 ***	0.7 ± 0.1 **	49.7 ± 0.7 **	0.9 ± 0.1 **	51.3 ± 0.6 ***
line 12	4.3 ± 0.2	28.4 ± 1.4 ***	16.0 ± 0.7 ***	3.0 ± 0.2 ***	0.4 ± 0.0	47.9 ± 0.7 **	0.0 ± 0.0 ***	48.3 ± 0.7 ***
line 13	4.4 ± 0.1	22.7 ± 2.8 *	17.4 ± 0.6 ***	1.9 ± 0.4 ***	0.4 ± 0.1	53.2 ± 2.0	0.0 ± 0.0 ***	53.6 ± 2.0
line 14	4.3 ± 0.3	24.3 ± 2.0 **	14.6 ± 1.1 ***	4.5 ± 0.9 ***	0.4 ± 0.0	50.9 ± 1.1*	1.1 ± 0.5 **	52.4 ± 1.6 **
line 15	4.0 ± 0.0 ***	26.3 ± 1.6 ***	14.5 ± 0.4 ***	4.0 ± 0.6 ***	0.4 ± 0.0	50.1 ± 0.9 *	0.7 ± 0.2 ***	51.2 ± 1.0 **
average of transgenic line	4.4 ± 0.4	27.8 ± 5.9	14.6 ± 1.7	4.4 ± 1.6	4.7 ± 4.5	43.5 ± 7.2	0.7 ± 0.4	48.9 ± 5.0

Three or four replicates of 30-seed sample were measured for wild-type and each transgenic line. All data are averages of three measurements ±SD. Fatty acid legend: 18:1 is oleic; 18:2 is linoleic; 18:3 is linolenic; 18:1OH is ricinoleic; 20:1OH is lesquerolic; and 20:2OH is auricolic acid. a, total content of five common fatty acids: palmitic (16:0), palmitoleic (16:1), stearic (18:0), arachidic (20:0), and eicosenoic acids (20:1). Two-tailed Student's t-test. *$^{*}p < 0.05$.*

$^{**}p < 0.01$.

$^{***}p < 0.001$.

Table 2.

Fatty acid composition (mole %) in T_1 seeds expressing CsFAD2 RNAi + AtFAD3 RNAi + CsFAE1 RNAi.

We have demonstrated that high levels of 18:1OH can be achieved by blocking the elongation of 18:1OH to 20:1OH. Also, high levels of 18:1 and 18:2 were accumulated through suppression of desaturation steps. However, the accumulated 18:1 was not converted to 18:1OH and instead, 18:1 was largely channeled to seed TAG.

4. Bottlenecks and potential for production of a high 18:1OH-containing oil in lesquerella

Based on the data presented in **Tables 1** and **2**, significant amount of 18:1 was not utilized for 18:1OH production. One of the factors could be due to the bifunctional activity of PfFAH12 [8], which hydroxylates and desaturases 18:1 to produce 18:1OH and 18:2, respectively, thus diverting 18:1 flux to 18:2 production. Seeds contain 90% 18:1OH from castor [37] or 85% 20:1OH from *Physaria lindheimeri* [38, 39]. These species have strict *FAH12*s, *RcFAH12* [7] and *PlFAH12* [38]. Deleting *PfFHA12* and at the same time introducing *RcFAH12* or *PlFAH12* in lesquerella should allow increased 18:1OH accumulation. On the other hand, the accumulation of 18:1 could also be due to a lesquerella LPAT that has substrate preference for 18:1-CoA, resulting in efficient incorporation of 18:1-CoA into TAG through Kennedy pathway. We already showed that castor *RcLPAT2* increased 18:1OH in lesquerella [20, 21]. Besides, castor *RcLPAT3B* and *RcLPATB* also showed substrate preference to 18:1OH in Arabidopsis [40]. To enhance 18:1OH to a higher level in lesquerella, further engineering design should include knocking out a lesquerella *PfLPAT2* and overexpressing *RcLPAT2*, *RcLPATB*, and *RcLPAT3B*. In addition to Kennedy pathway, some of the 18:1-PC could be converted by PDCT to 18:1-DAG for TAG assembly in lesquerella (**Figure 1**). Lesquerella seed TAGs contain about 21% 18:2 and 18:3 (**Tables 1** and **2**). There is strong evidence that seeds enriched with 18:2 or 18:3 may use the PC-derived pathway [2]. Therefore, it is likely that PC-derived DAGs are utilized in TAG assembly in lesquerella. To enhance flux from 18:1OH-PC to 18:1OH-DAG, it is advantageous to replace a lesquerella *PfPDCT* with a castor *RcPDCT* which was demonstrated effective in converting 18:1OH-PC to 18:1OH-DAG in Arabidopsis [25].

We observed increase in 18:1OH and decrease in 20:1OH in transgenic lesquerella seeds expressing *CsFAE1 RNAi*, which are expected. We, however, found that total HFAs are reduced (**Tables 1** and **2**). Considering lesquerella PfKCS18 is evolved to specifically elongate 18:1OH-CoA to 20:1OH-CoA [10] (**Figure 1**), it is possible that some enzymes in Kennedy pathway are co-evolved to adapt and utilize 20:1OH efficiently. Introducing additional enzymes with 18:1OH substrate selectivity may enhance total HFA level in lesquerella seeds. Although most plant GPATs select a wide range of acyl-CoA substrates [2, 41], castor RcGPAT9 was able to incorporate HFAs including 18:1OH at the *sn*-1 position of G3P, thus playing a critical role for *sn*-2 and *sn*-3 HFA acylation by LPAT and DGAT [42, 43]. Castor RcDGAT2 prefers 18:1OH to common FAs for esterifying 18:1OH to the *sn*-3 position of DAG [44, 45]. Thus future engineering design may target *RcGPAT9* and *RcDGAT2*. Lesquerella seed transcriptome analysis reveals one PfGPAT9 and three PfDGATs [46]. Evaluation on substrate preference by these genes will provide insights for enzyme characteristics in HFA-rich species.

One of the engineering examples demonstrates the interplay between Kennedy pathway and PC-mediated pathway for acylating HFA into *PfKCS18* was found to efficiently elongate 18:1OH to 20:1OH [47] in transgenic camelina expressing *RcFAH12* [32]. The transgenic camelina seeds expressing both *RcFAH12* with *PfKCS18* increased HFA content to 21% compared with the background line expressing single *RcFAH12* [47]. 18:1OH-PC generated by RcFAH12 in camelina may be subjected to β-oxidation [48], or represents a bottleneck [24], because camelina is not equipped with enzymes and pathways for channeling 18:1OH-PC into storage TAG. PfKCS18 may ease the 18:1OH flux from PC to cytosol by converting 18:1OH to 20:1OH, thus relieving the bottleneck and facilitating the incorporation of HFA into TAG by Kennedy pathway (**Figure 1**) [16].

FA at the *sn*-2 position of PC can be transferred to the *sn*-3 position of DAG, by PDAT [28, 49] (**Figure 1**). Castor *RcPDAT1–2* (or *RcPDAT1A*) transfers 18:1OH from its PC to DAG for HFA-TAG synthesis [29, 30]. To further enhance 18:1OH accumulation in lesquerella TAGs, *RcPDAT1–2* is another candidate target for genetic engineering.

5. Summary

Significant increases in 18:1OH content are achieved through over expressing *RcLPAT2* and silencing *FAD2*, *FAD3* and *KCS18*. Intriguingly, the accumulated 18:1 was not efficiently utilized to produce 18:1OH and instead, 18:1 was largely channeled to seed TAG. Future research efforts may focus on implementing genetic approach that targets not only enhancement of 18:1OH synthesis, but also on increased 18:1OH acylation to TAG. These genes include *RcFAH12* or *PlFAH12*, *RcGPAT9*, *RcLPAT2*, *RcDGTAT2*, *RcPDCT*, *RcLPAT1–2*. Nevertheless, we have demonstrated that lesquerella can be engineered for large increases in 18:1OH levels from 0.4–0.5% in WT to a stable high level of 15–20% in transgenic seed oils.

Conflicts of interest

The authors declare no conflict of interest.

Notes/thanks/other declarations

Authors note that **Figure 1** was revised from the original publication listed in ref. [36]; **Figures 2** and **3** are cited from the original publication listed in ref. [21]; **Tables 1** and **2** are cited from the original publication listed in ref. [36].

Author details

Grace Chen* and Kumiko Johnson
U.S. Department of Agriculture, Western Regional Research Center, Agricultural Research Service, Albany, CA, USA

*Address all correspondence to: grace.chen@usda.gov

References

[1] Li-Beisson Y et al. Acyl-Lipid metabolism. In: The Arabidopsis Book. Vol. 11. Rockville, MD: American Society of Plant Biologists; 2013. p. e0161

[2] Bates PD. Understanding the control of acyl flux through the lipid metabolic network of plant oil biosynthesis. Biochimica et Biophysica Acta (BBA) - Molecular and Cell Biology of Lipids. 2016;**1861**(9, Part B):1214-1225

[3] Lager I et al. Plant acyl-CoA: Lysophosphatidylcholine acyltransferases (LPCATs) have different specificities in their forward and reverse reactions. The Journal of Biological Chemistry. 2013;**288**(52):36902-36914

[4] Lands WE. Lipid Metabolism. Annual Review of Biochemistry. 1965;**34**:313-346

[5] Bafor M et al. Ricinoleic acid biosynthesis and triacylglycerol assembly in microsomal preparations from developing castor-bean (Ricinus communis) endosperm. Biochemical Journal. 1991;**280**(2):507-514

[6] Moreau RA, Stumpf PK. Recent studies of the Enzymic synthesis of Ricinoleic acid by developing Castor beans. Plant Physiology. 1981;**67**(4):672-676

[7] Van De Loo FJ et al. An oleate 12-hydroxylase from Ricinus communis L. is a fatty acyl desaturase homolog. Proceedings of the National Academy of Sciences of the United States of America. 1995;**92**(15):6743-6747

[8] Broun P, Boddupalli S, Somerville C. A bifunctional oleate 12-hydroxylase: Desaturase from Lesquerella fendleri. Plant Journal. 1998;**13**(2):201-210

[9] Bayon S et al. A small phospholipase A2-α from castor catalyzes the removal of hydroxy fatty acids from phosphatidylcholine in transgenic Arabidopsis seeds. Plant Physiology. 2015;**167**(4):1259-1270

[10] Moon H, Smith MA, Kunst L. A condensing enzyme from the seeds of Lesquerella fendleri that specifically elongates hydroxy fatty acids. Plant Physiology. 2001;**127**(4):1635-1643

[11] Okuley J et al. Arabidopsis FAD2 gene encodes the enzyme that is essential for polyunsaturated lipid synthesis. The Plant Cell. 1994;**6**(1):147

[12] Arondel V et al. Map-based cloning of a gene controlling omega-3 fatty acid desaturation in Arabidopsis. Science. 1992;**258**(5086):1353-1355

[13] Engeseth N, Stymne S. Desaturation of oxygenated fatty acids in Lesquerella and other oil seeds. Planta. 1996;**198**(2):238-245

[14] Reed DW, Taylor DC, Covello PS. Metabolism of hydroxy fatty acids in developing seeds in the genera Lesquerella (Brassicaceae) and Linum (Linaceae). Plant Physiology. 1997;**114**(1):63-68

[15] Lee K-R et al. Lesquerella FAD3-1 gene is responsible for the biosynthesis of trienoic acid and dienoic hydroxy fatty acids in seed oil. Industrial Crops and Products. 2019;**134**:257-264

[16] Kennedy EP. Biosynthesis of complex lipids. Federation Proceedings. 1961;**20**:934-940

[17] Hayes DG, Kleiman R. 1,3-specific lipolysis of lesquerella fendleri oil by

immobilized and reverse-micellar encapsulated enzymes. JAOCS, Journal of the American Oil Chemists' Society. 1993;**70**(11):1121-1127

[18] Lin J-T, Chen GQ. Quantification of the molecular species of TAG and DAG in Lesquerella (Physaria fendleri) oil by HPLC and MS. Journal of the American Oil Chemists' Society. 2014;**91**(8):1417-1424

[19] Lin J-T, Fagerquist CK, Chen GQ. Ratios of Regioisomers of the molecular species of Triacylglycerols in Lesquerella (Physaria fendleri) oil estimated by mass spectrometry. Journal of the American Oil Chemists' Society. 2016;**93**(2):183-191

[20] Chen GQ et al. Regiobiochemical analysis reveals the role of castor LPAT2 in the accumulation of hydroxy fatty acids in transgenic lesquerella seeds. Biocatalysis and Agricultural Biotechnology. 2020;**25**:101617

[21] Chen GQ et al. Expression of castor LPAT2 enhances ricinoleic acid content at the sn-2 position of triacylglycerols in lesquerella seed. International Journal of Molecular Sciences. 2016;**17**(4):507

[22] Frentzen M. Acyltransferases from basic science to modified seed oils. Fett. 1998;**100**(4-5):161-166

[23] Lu C et al. An enzyme regulating triacylglycerol composition is encoded by the ROD1 gene of Arabidopsis. Proceedings of the National Academy of Sciences of the United States of America. 2009;**106**(44):18837-18842

[24] Bates PD, Browse J. The pathway of triacylglycerol synthesis through phosphatidylcholine in Arabidopsis produces a bottleneck for the accumulation of unusual fatty acids in transgenic seeds. Plant Journal. 2011;**68**:387-399

[25] Hu Z, Ren Z, Lu C. The phosphatidylcholine diacylglycerol cholinephosphotransferase is required for efficient hydroxy fatty acid accumulation in transgenic Arabidopsis. Plant Physiology. 2012;**158**(4):1944-1954

[26] Slack CR et al. Some evidence for the reversibility of the cholinephosphotransferasecatalysed reaction in developing linseed cotyledons in vivo. Biochimica et Biophysica Acta (BBA)/lipids and lipid. Metabolism. 1983;**754**(1):10-20

[27] Aryal N, Lu C. A phospholipase C-like protein from Ricinus communis increases Hydroxy fatty acids accumulation in transgenic seeds of Camelina sativa. Frontiers in Plant Science. 2018;**9**:1576

[28] Dahlqvist A et al. Phospholipid: Diacylglycerol acyltransferase: An enzyme that catalyzes the acyl-CoA-independent formation of triacylglycerol in yeast and plants. Proceedings of the National Academy of Sciences of the United States of America. 2000;**97**(12):6487-6492

[29] Kim HU et al. Endoplasmic reticulum-located PDAT1-2 from castor bean enhances hydroxy fatty acid accumulation in transgenic plants. Plant and Cell Physiology. 2011;**52**(6):983-993

[30] van Erp H et al. Castor phospholipid: Diacylglycerol acyltransferase facilitates efficient metabolism of hydroxy fatty acids in transgenic Arabidopsis. Plant Physiology. 2011;**155**(2):683-693

[31] Lin JT, Chen GQ. Identification of TAG and DAG and their FA constituents in Lesquerella (Physaria fendleri) oil by HPLC and MS. JAOCS, Journal of the American Oil Chemists' Society. 2013;**90**(12):1819-1829

[32] Lu C, Kang J. Generation of transgenic plants of a potential oilseed crop Camelina sativa by agrobacterium-mediated transformation. Plant Cell Reports. 2008;**27**(2):273-278

[33] Nguyen HT, Silva JE, Podicheti R, Macrander J, Yang W, Nazarenus TJ, et al. Camelina seed transcriptome: A tool for meal and oil improvement and translational research. Plant Biotechnology Journal. 2013;**11**:759-769. DOI: 10.1111/pbi.12068

[34] Kumssa TT, Nazarenus TJ, Koster KL, Nguyen HT, Cahoon RE, Lu C, et al. Field performance and enhanced oil oxidative stability of the biofuel and industrial oilseed Camelina engineered for reduced fatty acid polyunsaturation. 2021 [in preparation]

[35] Chen GQ et al. Genetic engineering of Lesquerella with increased Ricinoleic acid content in seed oil. Plants. 2021;**10**(6):1093

[36] Chen GQ, Lin JT, Lu C. Hydroxy fatty acid synthesis and lipid gene expression during seed development in Lesquerella fendleri. Industrial Crops and Products. 2011;**34**(2):1286-1292

[37] Chen GQ et al. Expression profiles of genes involved in fatty acid and triacylglycerol synthesis in castor bean (Ricinus communis L.). Lipids. 2007;**42**(3):263-274

[38] Dauk M et al. A FAD2 homologue from Lesquerella lindheimeri has predominantly fatty acid hydroxylase activity. Plant Science. 2007;**173**(1):43-49

[39] Chen GQ et al. Seed development and hydroxy fatty acid biosynthesis in Physaria lindheimeri. Industrial Crops and Products. 2017;**108**:410-415

[40] Kim HU et al. Variant Castor lysophosphatidic acid acyltransferases Acylate Ricinoleic acid in seed oil. Industrial Crops and Products. 2020;**150**:112245

[41] Waschburger E et al. Genome-wide analysis of the Glycerol-3-phosphate acyltransferase (GPAT) gene family reveals the evolution and diversification of plant GPATs. Genetics and Molecular Biology. 2018;**41**(1 suppl. 1):355-370

[42] Lunn D, Wallis JG, Browse J. Tri-Hydroxy-triacylglycerol is efficiently produced by position-specific Castor acyltransferases. Plant Physiology. 2019;**179**(3):1050-1063

[43] Shockey J et al. Specialized lysophosphatidic acid acyltransferases contribute to unusual fatty acid accumulation in exotic Euphorbiaceae seed oils. Planta. 2019;**249**(5):1285-1299

[44] Kroon JTM et al. Identification and functional expression of a type 2 acyl-CoA:Diacylglycerol acyltransferase (DGAT2) in developing castor bean seeds which has high homology to the major triglyceride biosynthetic enzyme of fungi and animals. Phytochemistry. 2006;**67**(23):2541-2549

[45] Burgal J et al. Metabolic engineering of hydroxy fatty acid production in plants: RcDGAT2 drives dramatic increases in ricinoleate levels in seed oil. Plant Biotechnology Journal. 2008;**6**(8):819-831

[46] Kim HU, Chen GQ. Identification of hydroxy fatty acid and triacylglycerol metabolism-related genes in lesquerella through seed transcriptome analysis. BMC Genomics. 2015;**16**:230

[47] Snapp AR et al. A fatty acid condensing enzyme from Physaria

fendleri increases hydroxy fatty acid accumulation in transgenic oilseeds of Camelina sativa. Planta. 2014;**240**(3):599-610

[48] Moire L et al. Impact of unusual fatty acid synthesis on futile cycling through beta-oxidation and on gene expression in transgenic plants. Plant Physiology. 2004;**134**(1):432-442

[49] Stahl U et al. Cloning and functional characterization of a phospholipid:Diacylglycerol acyltransferase from Arabidopsis. Plant Physiology. 2004;**135**(3):1324-1335

Section 2

Omega-3 Fatty Acids and Cardiovascular Disease

Chapter 3

N-3 Polyunsaturated Fatty Acids and Their Role on Cardiovascular System

Savina Nodari and Francesco Fioretti

Abstract

The interest in n-3 polyunsaturated fatty acids (n-3 PUFAs), their favorable effects on the cardiovascular (CV) risk profile and prevention of CV events has been growing over the years, leading to their recommendation for secondary prevention in post myocardial infarction and hypertriglyceridemia. However, years later conflicting results provided by clinical trials have generated some doubts about their CV benefits, leading to a limited indication for the treatment of hypertriglyceridemia. Only recently, after the REDUCE-IT Trial results on CV events and mortality, n-3 PUFAs have recovered an indication in the international guidelines for hypertriglyceridemia in patients with Atherosclerotic Cardiovascular Disease (ASCVD) or with type 2 diabetes mellitus (T2DM) and other CV risk factors, already on statin therapy. Multiple beneficial CV effects have been highlighted, in addition to the well-known lipid-lowering function, such as anti-inflammatory, anti-thrombotic and endothelial function protective properties. Three formulations of n-3 PUFAs are currently available on the market, sharing some pharmacokinetic and pharmacodynamic characteristics, but also exhibiting peculiar mechanisms. Three major clinical trials evaluated the efficacy and safety of different formulations of n-3 PUFA: JELIS, REDUCE-IT and STRENGTH, with controversial results attributable to various factors. For the future, it could be useful to perform comparative studies between different formulations and placebo, in order to clarify these doubts.

Keywords: n-3 PUFAs, cardiovascular diseases, cardiovascular prevention, coronary artery disease, IPE

1. Introduction

Fatty acids (FA) have the typical RCOOH structure, containing a methyl group, a hydrocarbon chain (R) and a carboxyl group (**Figure 1**). The carbon chain (C) can vary greatly in length, with a range between 2 and 36 C atoms. They are divided into saturated, monounsaturated and polyunsaturated according to the number of double bonds C = C present in their molecule (respectively none, one or more than one). FA have both a systematic and common name (e.g., octadecanoic or stearic acid) [2] and are often expressed with a schematic formula (abbreviated notation): CN: p n-x,

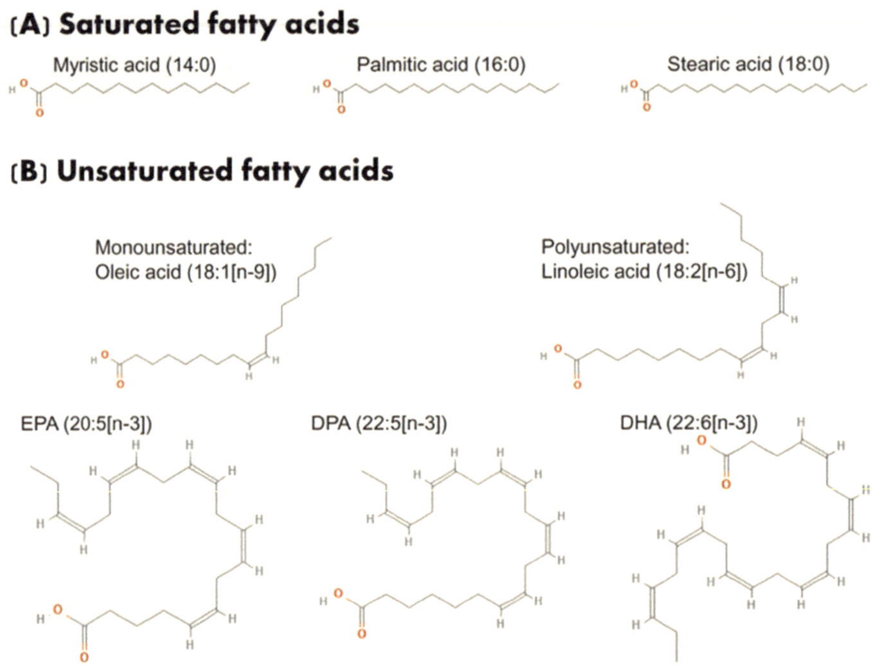

Figure 1.
Chemical structure of key saturated (A) and unsaturated (B) fatty acids. Modified from Wu H et al. [1]. DHA, docosahexaenoic acid; DPA, docosapentaenoic acid; EPA, eicosapentaenoic acid.

where CN represents the total number of atoms of C, p the number of double bonds, x the position of the first double bond from the methyl end (n) [3].

Polyunsaturated (PUFA) n-3 (Omega 3, PUFA n-3) and n-6 (Omega 6, PUFA n-6 Fatty Acids), whose first double bond is respectively on the third and sixth atom of C (numbered by the methyl group of the carbon chain), have a particular biological and medical interest [3].

Linoleic (AL) and alpha-linolenic (ALA) acids are defined as "essentials," because they cannot be synthesized *de novo* in humans and must, therefore, be taken with the diet (**Table 1**). Moreover, they are considered the precursors of the n-3 and n-6 PUFA families [4].

PUFA n-3. They include ALA (18,3 n-3) of plant origin and FA of animal origin (in particular seafood—fish oil), mainly represented by eicosapentaenoic acids (EPA; 20:5 n-3), docosahexaenoic acids (DHA; 22:6n-3) and docosapentaenoic acids (DPA, 22:5 n-3). It has been shown that ALA induces CV benefits in observational studies [5] and fish intake protects against coronary heart disease risk [6]. Moreover, a negative correlation between n-3 PUFAs and CV mortality has been reported in clinical trials and meta-analyses [7].

PUFA n-6. The main n-6 PUFAs are AL and arachidonic acid (AA). AL is the main n-6 PUFA derived by food and is contained in vegetable oils, nuts and seeds; AA is mainly taken from red meat, eggs, seaweed and fish oil [1]. A high intake of n-6 PUFAs, predominantly AL, would appear to be associated with a reduced risk of ischemic heart disease, ischemic stroke and CV mortality [8].

Several studies have shown the importance to maintain the balance between n-6 and n-3 PUFAs rather than the absolute quantity of each individual molecule. The centrality of this balance has been emphasized not only in the cardiology context, but

Product	LA	ALA	AA	EPA + DHA
$n-6$ FA rich foods				
Corn oil	50,000	900		
Cotton seed oil	47,800	1000		
Peanut oil	23,900			
Soybean oil	53,400	7600		
Sunflower oil	60,200	500		
Safflower oil	74,000	470		
Margarine	17,600	1900		
Lard	8600	1000	1070	
Chicken egg	3800	220		
Bacon	6080	250	250	
Ham	2480	160	130	
Soya bean	8650	1000		
Maize	1630	40		
Almond	9860	260		
Brazil nut	24,900			
Peanut	13,900	530		
Walnut	34,100	6800	590	
$n-3$ FA rich foods				
Canola oil	19,100	8600		
Linseed oil	13,400	55,300		
Herring	150	61.66	36.66	1700
Salmon	440	550	300	1200
Trout	74		30	500
Tuna	260	270	280	400
Cod	4	2	3	300

[a]*Data reported as mg/100 g.*
[b]*Content of* $n-6$, $n-3$ *FAs may slightly vary according to species, sources and analytical methods. Modified from Russo GL et al. [10]. LA, Linoleic Acid; ALA, Alfa-linolenic acid; AA, Arachidonic acid; EPA, eicosapentaenoic acid; DHA, docosahexaenoic acid.*

Table 1.
Dietary sources of principal polyunsaturated fatty acids (PUFAs)[a, b].

also in the pathogenesis of oncological, inflammatory and autoimmune diseases. The pleiotropic effects of PUFAs are summarized in **Table 2** [9].

For example, a high n-6/n-3 ratio is considered harmful to human health, while a value close to 1 is considered protective against degenerative diseases [10].

In healthy subjects taken a typical Swedish diet, a serum PUFA n-6/n-3 ratio of 4.72:1 was observed to be associated with an increase in the number of leukocytes and platelets and VEGF levels, when compared with an n-6/n-3 PUFA ratio of 2.6:1, observed in individuals following a Mediterranean diet. This dietary comparison and

Antithrombotic and antiplatelet
Decreased platelet aggregation: EPA decreased platelet aggregation and inhibited AA-induced thromboxane A2 formation
Decreased platelet adhesion: • EPA reduced platelet adhesion to collagen type I by 60–65% • Fish oil equivalent to 6 g EPA/day reduced platelet pseudopodia
Decreased thromboxane A2/increased thromboxane A3 levels: • EPA added to human thrombocytes incubated with AA decreased thromboxane A2 and increased thromboxane A3 • In human platelet-rich plasma, EPA did not induce platelet aggregation
Anti-inflammatory
Decreased hsCRP: • In ANCHOR, IPE 4 g/day decreased hsCRP vs. placebo by 21.5% ($P < 0.01$) • In REDUCE-IT, IPE decreased hsCRP vs. placebo at year 2 by 39.9% (last visit, 37.6%); $P < 0.001$
Decreased pentraxin-3: In patients with CAD treated with PCI, EPA added to statin therapy after 9 months reduced pentraxin 3 levels vs. statin alone
Decreased Lp-PLA$_2$: In ANCHOR, IPE 4 g/day decreased Lp-PLA$_2$ by 19.6% in patients with hsCRP >2.0 vs. placebo
Decreased NLRP3 inflammasome activation: • EPA blocked NLRP3 inflammasome activation in an animal model of ischemic stroke • EPA decreased NLRP3 gene expression in adipose tissue and in classically activated THP-macrophages
Decreased toll-like receptor 4: EPA reduced gene expression of toll-like receptor 4 by >50% in mice fed a high-fat diet vs. control
Decreased NFKb: EPA reduced the genetic expression of NFKb in a cell culture with THP-1 macrophages
Decreased inflammatory cytokines: • EPA reduced gene expression of IL-1b, TNF-α, and MCP-1 in cell culture with THP-1 macrophages • EPA downregulated expression of IL-6 mRNA in IL-1b–stimulated C6 glioma cells
Increased anti-inflammatory cytokine IL-10: • EPA significantly increased the production of IL-10 in lipopolysaccharide-activated monocytes ($P < 0.05$) • EPA increased IL-10 expression in peripheral blood monocytes in obese patients with dyslipidemia
Decreased endothelial adhesion molecules: • EPA inhibits monocyte adhesion to endothelial cells • EPA decreases plasma concentrations of soluble ICAM and VCAM
Increased "resolution of inflammation": specialized pro-resolving mediators
RvE1 was found to bind to the LTB4 receptor, BLT1, on human PMNs and attenuated LTB4-induced proinflammatory signals and PMN migration in an *in vitro* study
RvE2 was found to enhance phagocytosis of human macrophages and anti-inflammatory cytokine (IL-10) production
RvE3 was found to have potent inhibitory action on neutrophil chemotaxis both *in vitro* and *in vivo*
Anti-oxidant
Increased paraoxonase in patients with type 2 DM: • EPA 2 g increased PON1 activity and levels over 8 weeks vs. placebo • EPA over 8 weeks significantly increased gene expression of PON2 vs. placebo
Inhibition of lipid/lipoprotein oxidation: • EPA inhibited the oxidation of apoB lipoproteins (LDL-C, sdLDL, VLDL-C) in an *in vitro* study • EPA inhibited the oxidation of HDL isolated from the plasma of healthy volunteers; the sustained antioxidant effects of EPA on HDL oxidation were not replicated by DHA • EPA inhibited oxidation of sdLDL, model membranes, and cholesterol crystal domain formation

Improved endothelial function
EPA significantly increased NO production and reduced glucose inhibition of NO in cultured human endothelial cells
EPA significantly improved FMD by 51% in patients with elevated TG ($P < 0.0001$); EPA/AA ratio was found to be significantly associated with the change in FMD ($P = 0.010$)
EPA improved PFBF during reactive hyperemia to the level in normolipidemic controls, recovery of PFBF correlated positively with EPA levels ($P < 0.05$) and the EPA/AA ratio ($P < 0.01$)
EPA improved vascular function as measured by strain gauge plethysmography in patients with type 2 DM on statin therapy
EPA improved endothelial function in patients with CAD, LDL-C < 100 mg/dL on statin therapy, and impaired FMD
EPA with an atorvastatin metabolite improved endothelial function through increased NO in an *in vitro* study
Increased cholesterol efflux
Increasing EPA phosphatidylcholine content of reconstituted HDL-C was demonstrated to increase cholesterol efflux
EPA inhibited glucose-induced membrane cholesterol crystalline domain formation in an *in vitro* study using multilamellar vesicles
EPA and DPA inhibited oxidation of membrane cholesterol domains in an *in vitro* study utilizing multilamellar vesicles
Increased adiponectin
EPA after 3 months significantly increased adiponectin vs. placebo among obese patients with dyslipidemia ($P < 0.01$)
Change in pulse wave velocity, a measure of arterial stiffness, was negatively correlated with change in adiponectin ($P < 0.01$) and in IL-10 expression of monocytes ($P < 0.05$)
Anti-arrhythmic
EPA may reduce the risk of ventricular arrhythmia by inhibiting the fast sodium current (I_{Na}), L-type calcium inward current (I_{Ca}) and enhancing the slowly activating delayed rectifying outward potassium current (I_{KS}), thus shortening cardiac action potential
EPA reduces cytosolic Ca^{2+} overload, thus reducing the risk of triggered induced arrhythmia

Modified from Trivedi K et al. [11]. AA, arachidonic acid; apo, apolipoprotein; BLT1, leukotriene B4 receptor BLT1; Ca2+, calcium; CAD, coronary artery disease; DHA, docosahexaenoic acid; DPA, docosapentaenoic acid; DM, diabetes mellitus; EPA, eicosapentaenoic acid; FMD, flow-mediated dilation; hsCRP, high-sensitivity C-reactive protein; HDL-C, high-density lipoprotein cholesterol; ICAM, intercellular adhesion molecule; IL, interleukin; IPE, icosapent ethyl; LTB4, leukotriene B4 receptor 1; LDL-C, low-density lipoprotein cholesterol; Lp-PLA2: lipoprotein- associated phospholipase A2; MCP, monocyte chemoattractant protein; mRNA, messenger ribonucleic acid; NFKb, nuclear factor kappa-light-chain-enhancer of activated B cells; NLRP3, NLR family pyrin domain containing 3; NO, nitric oxide; PCI, percutaneous coronary intervention; PFBF, peak forearm blood flow; PON, paraoxonase; PMNs, polymorphonuclear leukocytes; REDUCE-IT, Reduction of Cardiovascular Events with Icosapent Ethyl–Intervention Trial; Rv, resolvin; sdLDL, small dense LDL; THP-1, monocyte/macrophage; TNF-α, tumor necrosis factor alpha; VLDL-C, very-low-density lipoprotein cholesterol; VCAM, vascular cell adhesion molecule.

Table 2.
Select pleiotropic effects of EPA.

the respective clinical implications are the basis for the recognition of the importance of the Mediterranean diet in maintaining a better state of health, enough to be included in the list of Intangible Heritage of Humanity of UNESCO [11].

2. Molecular mechanisms of n-3 PUFAs

N-3 PUFAs produce multiple effects, as anti-inflammatory, anti-thrombotic properties and actions on endothelial function, in addition to the well-known lipid lowering and hypotriglyceridemic effects (**Figure 2**).

The molecular mechanisms by which n-3 PUFAs exert their pleiotropic activity are complex and not yet fully defined.

Firstly, n-3 PUFAs are embedded in the cell wall and organelles, altering the lipid environment of membranes and their fluidity. This involves n-3 PUFAs in various cellular functions, such as signal transduction and protein transport. The beneficial effects of n-3 PUFAs observed in patients with hypertriglyceridemia seem to derive from their specific actions on inflammation, platelet function and monocyte phenotypes.

In particular, the production of specialized mediators, such as E-series resolvins derived from EPA and protectins and D-series resolvins derived from DHA, can play an important role in the resolution of inflammation. The reduction of the inflammatory response by n-3 PUFAs contributes to inhibit the development of intracellular second messengers, such as diacylglycerol and ceramide. In addition, n-3 PUFAs bind and activate different nuclear receptors and transcriptional factors involved in regulating the expression of different genes. These genes participate in functions related to

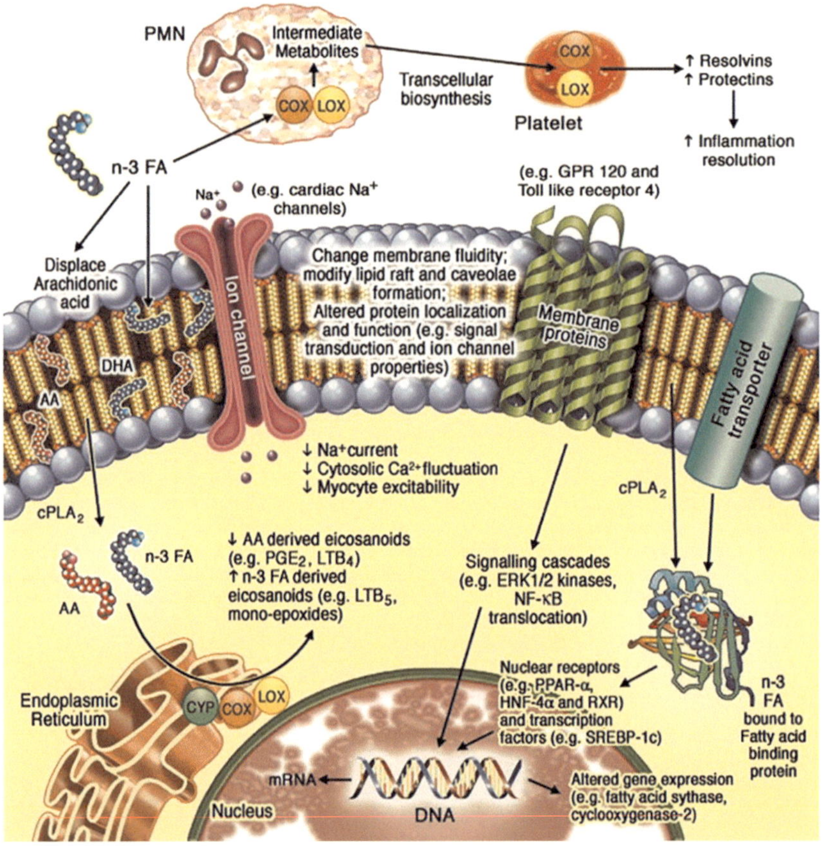

Figure 2.
Molecular pathways affected by n-3 PUFA. Taken from Mozaffarian D et al. [12]. DNA = deoxyribonucleic acid; ERK = extracellular signal-regulated kinase; mRNA = messenger ribonucleic acid; PMN = polymorphonuclear leukocyte.

atherosclerosis, such as inflammation, glucose-insulin homeostasis, lipid metabolism and adipocytokine production.

Finally, the activation of the peroxisome proliferator-activated receptor-γ by n-3 PUFAs results in the reduction of the translocation of NF-κB in the nucleus, decreasing the generation of multiple inflammatory cytokines [1].

3. Mechanisms underlying hypolipemic effect of n-3 PUFAs

Although the effect of n-3 PUFA-based drugs to reduce plasma triglycerides (TG) is widely recognized, their mechanism of action is not fully understood. The results of preclinical and clinical studies suggest that n-3 PUFAs reduce plasma TG concentrations by limiting their synthesis, but also by decreasing their incorporation into very low density lipoproteins (VLDLs), reducing their secretion and improving the clearance of TG by VLDLs [13, 14].

It has been proposed that n-3 PUFAs exert these effects through several mechanisms (**Figure 3**):

1. decrease in hepatic lipogenesis through suppression of the expression of sterol regulatory element-binding protein-1c (SREBP-1c). This leads to a decrease in the expression of enzymes responsible for the synthesis of cholesterol, fatty acids and TG;

2. increase in the β-oxidation of FA, with consequent reduction in the availability of the substrate necessary for the synthesis of TG and VLDL [13];

3. inhibition of key enzymes involved in hepatic TG synthesis, such as acid phosphatidic phosphatase and diacylglycerol acyltransferase [16];

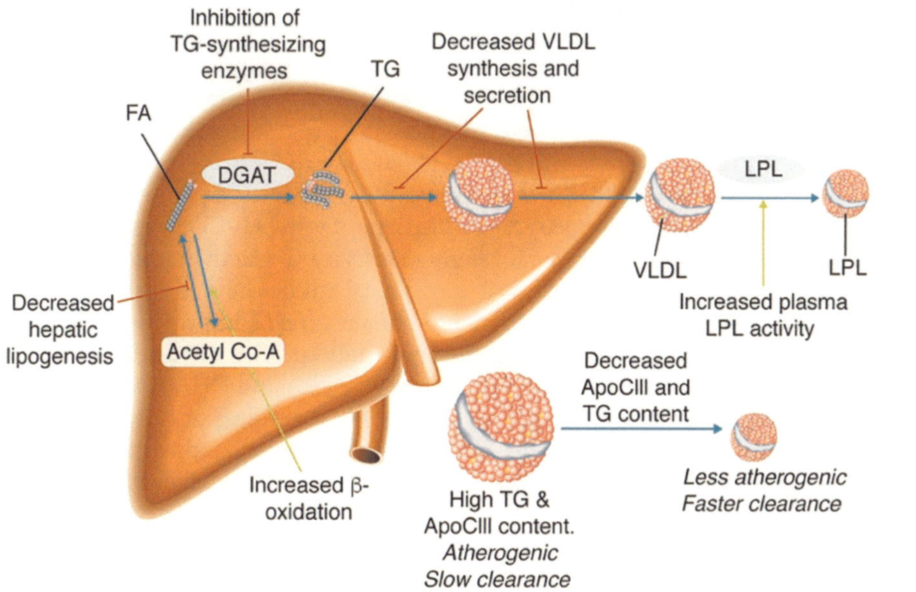

Figure 3.
Proposed mechanisms of action of prescription formulations of long-chain omega-3 fatty acids. Taken from Backes J et al. [15]. ApoCIII apolipoprotein CIII, acetyl Co-a acetyl coenzyme a, DGAT diglyceride acyltransferase; FAfatty acid, LPL lipoprotein lipase, TG triglyceride, VLDL very-low-density lipoprotein.

4. increase the expression of lipoprotein lipase (LPL), a key component of the biosynthetic pathways of TG-rich lipoproteins (TRLs), resulting in greater removal of TG by VLDLs and circulating chylomicrons.

The primary constituents of n-3 PUFA formulations, EPA and DHA, have both been shown to reduce TG. However, the two n-3 PUFAs are known to produce different effects on LDL-C and HDL-C [17]. In a meta-analysis of studies in which the effects of DHA and EPA were directly compared, DHA was associated with a greater reduction in TG, but also with a greater increase in LDL-C compared to EPA. In addition, DHA was associated with an increase in HDL-C compared to placebo, but not compared to EPA. Further studies are needed to clarify the mechanisms and significance of these differences [17].

N-3 PUFAs have also been shown to reduce circulating levels of apolipoprotein CIII (ApoCIII), believed to be involved in the progression of atherosclerosis and CV diseases (CVDs) through multiple mechanisms, such as activation and potentiation of pro-inflammatory pathways. In addition, it inhibits the receptor-mediated absorption of TRLs and their remnants, slowing their clearance and promoting the formation of small dense LDL particles from VLDLs. ApoCIII also provides a key contribution to the pathogenesis of hypertriglyceridemia, mainly due to its inhibitory action on LPL that hydrolyzes plasma TRLs, producing free FA, chylomicron remnants and intermediate density lipoproteins (IDL). Some IDL particles may undergo further LPL-mediated hydrolysis, converting to LDL.

It has been hypothesized that the different effects of DHA and EPA on the lipid profile can be explained with their different interactions with ApoCIII synthesis. DHA is believed to reduce ApoCIII synthesis by regulating some hepatic transcriptional factors, such as hepatic nuclear factor-4-alpha (HNF4A) and forkhead box-O transcription factor O1 (FOXO1). In this context, DHA promotes more hydrolysis of VLDLs, resulting in greater conversion of VLDLs into LDL, ultimately resulting in the formation of larger, less dense LDL particles. Further studies are necessary to clarify these properties.

In addition to the TG-reducing effect, n-3 PUFAs have beneficial effects on cardiometabolic risk factors, having proved to significantly reduce markers of inflammation associated with atherosclerosis and the development of CVDs.

Clinical studies showed that levels of lipoprotein-associated phospholipase A2 (Lp-PLA2) are reduced by n-3 PUFAs and icosapent ethyl (IPE) compared to placebo, and that the concentration of high-sensitivity C-reactive protein (CRP) is decreased by IPE [18, 19]. Preclinical and clinical studies have also shown that EPA and DHA have antiarrhythmic and antioxidant effects, improving endothelial function and promoting a less atherogenic lipoprotein profile when administered in combination with statin therapy [20]. Moreover, EPA and DHA reduce biomarkers of platelet activity compared to placebo, regardless of concomitant intake of aspirin and statins, and this is supposed to be associated with inhibition of platelet aggregation. Regarding this latter property, DHA is believed to possess more intensive antiplatelet activity than EPA. In addition, DHA has been shown to reduce blood pressure and heart rate, and to play a protective effect on cognitive decline.

Although most studies on the therapeutic effects of n-3 PUFAs focused on EPA and DHA, a recent study investigated the mechanism of action of DPA. DPA levels were independently associated with a reduced risk of developing acute myocardial infarction (AMI) and coronary artery disease (CAD), while reduced circulating levels of

DPA were shown to be associated with the development of lipid-rich plaques and peripheral arterial disease (PAD).

EPA, DHA and DPA counteract platelet aggregation in a dose-dependent manner. However, DPA seemed to be the most intensive inhibitor of platelet activity, compared with EPA and DHA. As DHA and EPA, also DPA can reduce the expression of pro-inflammatory genes.

The mechanisms underlying the reduction of TG by long-chain n-3 PUFAs differ from those of other lipid-lowering drugs, such as statins. Therefore, if administered in combination, different lipid-lowering drugs could potentially have complementary beneficial effects on the lipid profile [13]. As proof of this, incremental reductions in TG levels were observed when n-3 PUFAs are administered with statins [21].

3.1 n-3 PUFAs and atheromatous plaques

The inverse relationship found between the intima-media thickness of the carotid artery and the treatment with n-3 PUFAs suggested their endothelium protective activity and anti-atherosclerosis effect. There is specific evidence that this result is achieved through a reduction in inflammation and increased plaque stability.

The relationship between the vulnerability of atheromatous plaque and the serum n-3/n-6 PUFA ratio was evaluated in a study that enrolled patients undergoing percutaneous coronary angioplasty [22]. The EPA/AA ratio was examined at the time of the patient's admission. An angioscopy examination of the culprit coronary lesion evaluated the color tone of the plaque (white 0; light yellow 1; yellow 2; intense yellow 3) and the possible presence of a thrombus. Patients with stable angina (n = 38) were divided into 2 groups based on a reduced (<0.37; n = 19) or elevated (≥0.37; n = 19) EPA/AA ratio. In the group with reduced EPA/AA, the maximum yellow tone of the plaques was significantly higher than in patients with stable angina and high EPA/AA ratio. Moreover, the number of non-culprit yellow plaques with a thrombus tended to be higher, without reaching statistical significance. Multivariate analysis revealed an association between a lower serum EPA levels with lower EPA/AA ratio and the finding of more vulnerable grade 3 yellow plaques.

In another trial, 49 thin-cap non-culprit atheromatous plaques (TCFA) were evaluated in 30 patients with untreated dyslipidemia [23]. Patients were randomized to EPA (1.8 g/day) + rosuvastatin (n = 15, 23 TCFA) or rosuvastatin alone (n = 15, 26 TCFA). Percutaneous interventions and a 9-month follow-up OCT were performed to assess morphological changes in TCFA. The EPA/AA ratio and pentraxin-3 (PTX3) levels were also evaluated.

Despite LDL-C levels during follow-up were comparable in the two groups, subjects treated with EPA + statin had a higher EPA/AA ratio and lower PTX3 levels than those receiving the statin alone. The OCT also showed that in the EPA + statin group there was a greater increase in the thickness of the fibrotic cap, a more evident reduction of the lipid arch and its length, as well as a lower macrophage accumulation. Such evidence suggested that concomitant use of EPA and rosuvastatin contributes to the stabilization of a vulnerable plaque compared with the use of statin alone and, presumably, through inhibition of arterial wall inflammation.

The recent randomized, double-blind controlled clinical trial EVAPORATE [24] evaluated the volume of atheromatous plaque, using multidetector CT (MDCT), after treatment with 4 g/day of IPE in combination with diet and statin therapy versus statin alone. The study enrolled 80 patients with coronary atherosclerosis documented by MDCT (one or more stenosis with narrowing >20%), already on statin therapy and with elevated TG levels (median value in both groups of 259.1 ± 78.1 mg/dL). MDCTs follow-up were performed after 9 and 18 months. The primary endpoint was low attenuation volume changes (LAP volume) at 18 months. In the IPE group the LAP plaque volume was reduced by 17%, while in the control group it was more than doubled (+109%; P = 0.0061). Significant differences were also observed in the progression index of other plaques, such as fibrous and fibro-adipose plaques, whose volumes decreased in the IPE group and progressed in the placebo group ($P < 0.01$).

4. Different pharmacological formulations (OM3EE, OM3CA, IPE, EPA/DPA/DHA EE)

4.1 General aspects

PUFA n-3 formulations are recommended, in combination with diet, for the reduction of serum TG in adults with hypertriglyceridemia.

Mainly three formulations of PUFA n-3 are currently available [13]:

- based on ethyl esters of long-chain fatty acids (OM3EE), composed mainly of EPA and DHA;
- CA-based of PUFA n-3 (OM3CA), composed of long-chain free FA, including EPA, DHA and DPA;
- IPE-based, composed of more than 96% of the purified and stabilized ethyl ester of EPA [12]. IPE is, therefore, a stable and highly purified compound containing a single active ingredient. It has been recently approved in the United States by the FDA for the prevention of Major CV events (MCV) in high-risk individuals (on statin therapy, with triglyceridemia >150 mg/dL, CV disease (CVD) or diabetes mellitus, and 2 or more CV risk factors). This approval is derived from numerous evidences. For example, the REDUCE-IT trial, in which a 25% reduction in the risk of CV events was observed in patients treated with IPE. In contrast, the other two formulations have a combination of n-3 PUFA products (e.g., DHA + EPA), with limited evidence in terms of CV event risk reduction [12]. Nevertheless, at approved doses all three commercially available formulations have been shown to be effective in reducing triglyceridemia and serum VLDL levels.

4.2 Bioavailability

The absorption of DHA and EPA has been shown to vary between formulations.

The OM3CA formulation has a bioavailability (area under the "plasma concentration-time" curve from zero to the last measured concentration value [$AUC_{(0-t)}$]) of fourfold greater for both EPA and DHA during low-fat consumption than observed with OM3EE [12]. Absorption of OM3EE and IPE requires pancreatic lipase-mediated hydrolysis. Since pancreatic lipase levels depend on both the amount and type of lipids ingested, the absorption of OM3EE is thought to depend strongly on the fat content of meals. In contrast, n-3 PUFAs in the free FA form, contained in OM3CA formulations, are not dependent on pancreatic lipase hydrolysis. Therefore, the bioavailability of EPA and DHA present in this formulation is less dependent on meal fat content than in OM3EE formulations. In one study (n = 54), conducted under low-fat intake conditions, 59% of subjects treated with OM3CA maintained an $AUC_{(0-t)}$ for EPA and DHA ≥50% compared to the $AUC_{(0-t)}$ observed under high-fat intake. In contrast, this finding was present in only 6% of OM3EE-treated subjects. This observation could be useful from a therapeutic point of view, as current guidelines recommend a very low fat diet in patients with severe hypertriglyceridemia [25].

4.3 Efficacy

All n-3 PUFA formulations, administered at a dose of 4 g/day, have been shown to significantly reduce TG, VLDL-C, non-C-HDL, and ApoB levels in hypertriglyceridemic and high CV risk patients [13]. Higher doses of n-3 PUFAs and higher baseline triglyceridemia levels are associated with greater percentage of reductions in TG [26]. This was also observed in the MARINE trial, in which different dosages of highly purified EPA ethyl ester (Vazkepa) were compared (**Figure 4**). This effect appears to be independent from the severity of hypertriglyceridemia.

In the JELIS study, the use of a formulation of highly purified EPA ethyl ester (Epadel) was shown to reduce CV events when combined with statin and a reduction in atheromatous plaque thickness. It has also demonstrated a good safety and tolerability profile.

In REDUCE-IT trial, on the other hand, 4 g/day of EPA ethyl ester was shown to result in favorable effects in terms of reducing ischemic events and CV mortality.

In contrast, in the STRENGTH trial, which used the OM3CA (EPA + DHA) formulation, a neutral effect in terms of CV outcome (MACE) was observed, thus not supporting the hypothesis of using n-3 PUFAs for CV event risk reduction.

DHA-containing formulations (OM3CA and OM3EE) may cause an increase in C-LDL levels, especially in patients with severe hypertriglyceridemia (TG >500 mg/dL). However, this is not associated with increased levels of non-HDL C or ApoB, which seems to be a better predictor of CV risk than C-LDL values [13]. This is particularly true for patients with hypertriglyceridemia, who often have low LDL-C levels [25].

Non-DHA-containing formulations (EPA and IPE) appear not to cause an increase in C-LDL levels [12], as shown in the MARINE [27] and ANCHOR [28] clinical trials, which studied efficacy and safety of IPE.

These different effects of EPA versus DHA on C-LDL levels can be attributed to one or more mechanisms:

- Firstly, DHA (but not EPA) may down-regulate C-LDL receptor (LDL-R) expression and LDL-R-mediated clearance by the liver, which is an important reverse regulator of LDL-C levels. This appears to occur through suppression of the expression of *sterol regulatory element binding protein-2* (SREBP-2) by DHA.

- Secondly, DHA may increase the activity of *cholesteryl ester transfer protein* (CETP), which promotes the transfer of lipid cores between lipoproteins, resulting in an increase of LDL-C and decrease of HDL-C levels. EPA, on the other hand, has a neutral effect on CETP activity.

- Finally, DHA appears to up-regulate lipoprotein lipase enzyme activity to a greater extent than EPA, thereby promoting the conversion of VLDL to LDL with consequent increase of LDL-C levels.

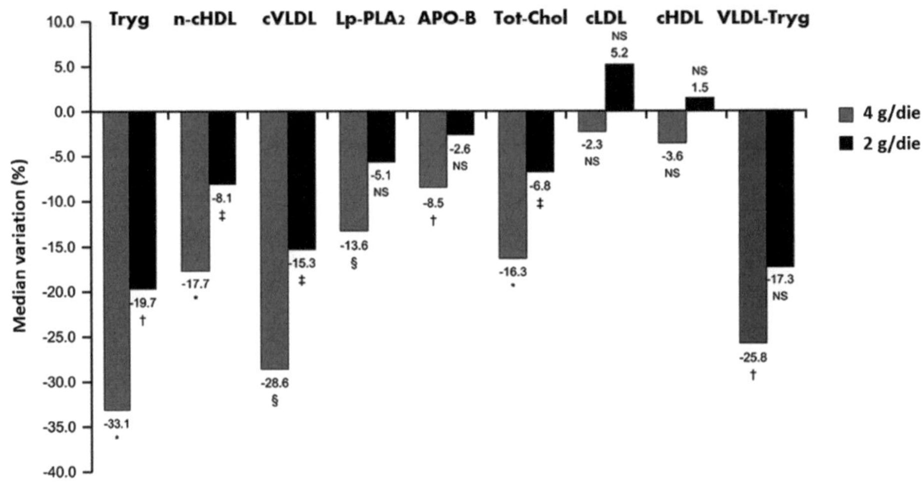

Figure 4.
Change in lipid levels, baseline to week 12 (intent-to-treat population). Percentage of change from baseline to study end of placebo-corrected lipid levels in patients with very high TG levels receiving AMR101 4 g/day or AMR101 2 g/day. Modified from Bays HE et al. [17]. APO-B = apolipoprotein B; cHDL = high-density lipoprotein cholesterol; cLDL = low-density lipoprotein cholesterol; Lp-PLA2 = lipoprotein associated phospholipase A2; n-cHDL = non-high-density lipoprotein cholesterol; NS = not significant; Tot-Chol = total cholesterol; Tryg = triglyceride; cVLDL, very-low-density lipoprotein cholesterol; VLDL–TG, very-low-density lipoprotein–triglycerides.

Interestingly, in patients with triglyceridemia >200 mg/dL but <500 mg/dL, no increase in LDL-C was observed with any of the formulations. This is probably because in the various clinical trials on n-3 PUFAs, such patients had lower baseline serum TG levels and were already on statin therapy.

It was also observed that, unlike IPE, DHA-containing formulations increased HDL-C levels. However, the clinical relevance of increasing HDL-C through pharmacological agents has not been clarified yet.

In contrast, to date, the effect of any n-3 PUFA formulation on pancreatitis has not been determined [13].

EPA and DHA are also believed to act in different or conflicting ways at the cell membrane level. Specifically, the insertion of EPA and DHA occurs in distinct regions of the lipid membrane bilayer as a consequence of the different lengths of the hydrocarbons involved. Their insertion causes conformational changes that result in increased membrane fluidity and promotion of cholesterol domains. Among the 2 n-3 PUFAs, EPA has a more stable and extended structure, which contributes to membrane stability as well as inhibition of lipid oxidation and formation of new cholesterol domains [29]. Their main effects are summarized in **Figure 5**.

4.4 Safety and tolerability

Commercially available n-3 PUFA formulations are generally well tolerated, with similar rates of treatment discontinuation between treatment and placebo groups [13]. In clinical trials, gastro-intestinal events were the main adverse effects with DHA-containing formulations, while arthralgia was the main adverse event reported with EPA [12].

In the REDUCE-IT trial, it was observed a higher incidence of atrial fibrillation and peripheral edema in the treatment group versus placebo group (5.3% vs. 3.9% and

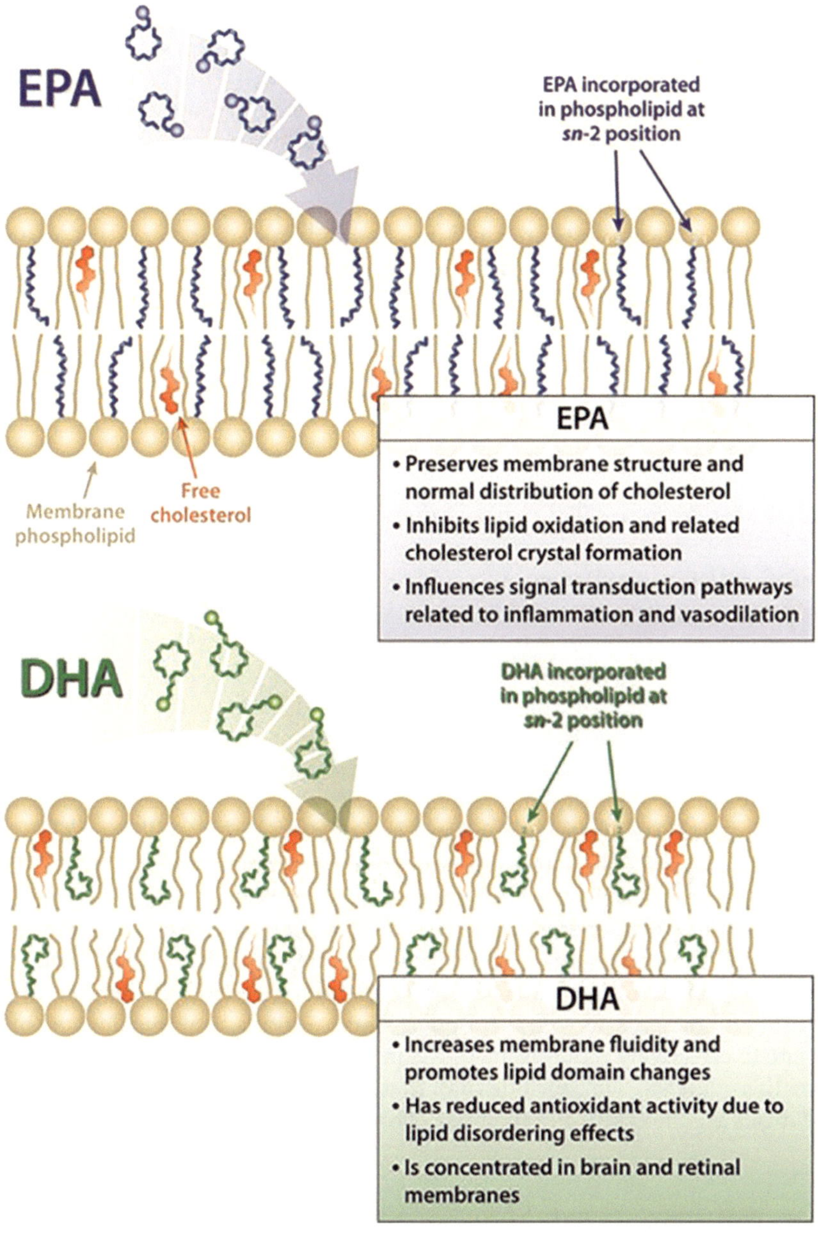

Figure 5.
Molecular membrane interactions of omega-3 fatty acids. Taken from Mason RP et al. [28].

6.5% vs. 5.0%, respectively). The possible pathogenetic mechanisms could be related to their effects on myocyte membrane ion channels. Previous observational and randomized studies have examined the potential antiarrhythmic effects of n-3 PUFAs, leading to conflicting results [30, 31]. However, in the REDUCE-IT trial, the increased incidence of atrial fibrillation in the EPA-treated group was not associated with a higher incidence of heart failure or stroke. Therefore, further studies are required to clarify these aspects.

In both JELIS and REDUCE-IT, modest increases in bleeding episodes were observed in the EPA-treated group compared with controls. Bleeding rates were lower in JELIS (1.1% in EPA vs. 0.6% in control group) than in REDUCE-IT (2.7% with EPA vs. 2.1% with placebo). In JELIS study antiplatelet therapy, at baseline, was present in 13% and 14% in treatment and placebo groups, respectively, compared with 79.7% and 79.1%, respectively, in the REDUCE-IT (26). Although no serious bleeding events were observed in REDUCE-IT or in other clinical trials with n-3 PUFAs, even in those who were on aspirin or warfarin therapy and in those undergone surgery or percutaneous interventions, the potential interaction between EPA and antithrombotic drugs warrants further investigation.

It is not known whether patients with allergies to fish and/or shellfish are at increased risk of allergic reactions to n-3 PUFAs. Therefore, these formulations should be used with caution in patients with known hypersensitivity to fish and/or shellfish [13].

4.5 Differences from n-3 PUFA-based dietary supplements

PUFA n-3 dietary supplements are widely used and they are among the most popular dietary supplements worldwide. However, they are not subject to the strict regulations required for prescription drugs. As a result, the EPA and DHA content of dietary supplements can be variable or too low [13].

Physicians often tend to erroneously consider n-3 PUFA dietary supplements to be adequate and reliable. However, a recent analysis of individual fish oil supplements has shown that they contain an inadequate dose of EPA and DHA, on average equivalent to only 68% of that required. The same analysis found that most supplements exceed the recommended levels of oxidation markers. The oxidative process undergone by n-3 PUFAs results in a gradual reduction in their concentration, leading to reduced efficacy. One study showed that a median intake of 11 servings of fish oil dietary supplements per day was needed to obtain a dose of 3.4 g/day of n-3 PUFAs. The same study showed that dietary supplements often contain other fats and cholesterol, and that their content varies widely among different products. Such variability can be confusing for patients, as well as for physicians, resulting in a drug dosage potentially inadequate to effectively reduce serum TGs in patients with hypertriglyceridemia [29]. Based on this evidence, we conclude that, to date, the benefits of n-3 PUFA-based nutraceuticals are uncertain and their use controversial [13].

5. Controlled clinical trials of different formulations of n-3 PUFAs. Efficacy on lipid profile and outcomes

The efficacy on lipid profile and outcomes of controlled clinical trials are summarized in **Figure 6** and **Table 3**, respectively. **GISSI-Prevention trial (1999)**. This multicenter [32] "open-label" study evaluated 11,324 patients with recent (<3 months) AMI randomized equally into four treatment groups: group 1, n-3 PUFAs (850–882 mg EPA and DHA as ethyl esters in EPA/DHA 1,2 ratio); group 2, vitamin E (300 mg, in 1 cps of synthetic α-tocopherol); group 3, n-3 PUFAs + vitamin E; control group 4 (no supplement). Combined primary efficacy endpoint represented by cumulative rate of all-cause mortality, nonfatal AMI, and nonfatal stroke; and cumulative rate of death from CV causes, nonfatal AMI, and nonfatal stroke.

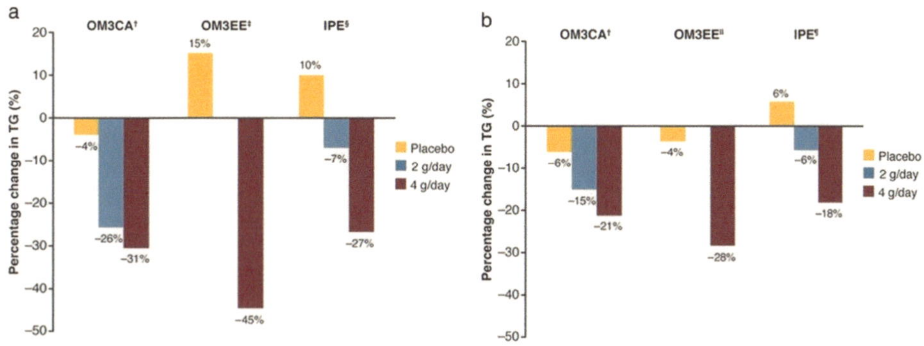

Figure 6.
TG reduction observed with varying doses of omega-3 carboxylic acids, omega-3 ethyl esters, and icosapent ethyl in (a) patients with severe hypertriglyceridemia (TG ≥ 500 mg/dL) and (b) statin-treated patients with high baseline TG levels (TG level ≥ 200 mg/dL and < 500 mg/dL). Taken from Backes J et al. [15].
OM3CA = percentage change from baseline expressed as least-squares geometric mean; OM3EE = percentage change from baseline expressed as geometric mean; IPE = percentage change from baseline expressed as median.
†Placebo used = olive oil; ‡placebo used = vegetable oil; §placebo used = mineral oil; ‖placebo used = corn oil; ¶Placebo not specified. IPE icosapent ethyl, OM3CA omega-3 carboxylic acids, OM3EE omega-3 ethyl esters, TG triglyceride.

Secondary analyses were performed for each component of the primary endpoints and for major causes of death.

Treatment with n-3 PUFAs, but not vitamin E, significantly reduced the primary endpoint (RR reduction of 10% (95% CI 1–18) with two-way analysis of variance and 15% (2–26) with four-way analysis). Benefits were attributed to reduction in risk of death (14% (3–24) with two-way, 20% (6–33) four-way) and CV death (17% (3–29) two-way, 30% (13–44) four-way). No significant change in C values (total, HDL, LDL) nor fibrinogen was observed in all groups.

In comparison with the control group, the slight reduction in triglyceridemia was shown more frequently in patients treated with PUFA n-3. The effects of combined treatment were similar to those observed in treatment with n-3 PUFAs alone, for both the primary endpoint (14% (1–26)) and fatal events (20% (5–33)).

These results demonstrate the statistically significant clinical benefits of dietary supplementation with n-3 PUFAs, where vitamin E supplementation demonstrated no benefit.

JELIS (Japan EPA Lipid Intervention Study; 2007). Randomized "open label" trial with blinded endpoint analysis [33] evaluated the long-term prevention effects of EPA of major coronary events in patients with hypercholesterolemia.

Over a 3-year period, 18,645 patients (60% women; mean age 61 years) with coronary artery disease (established angina pectoris, coronary interventions, previous AMI; n = 3664) or free of coronary artery disease (n = 14,981), with baseline total C value >250 mg/dL (corresponding to C-LDL >170 mg/dL) were enrolled. First-line therapy for all patients consisted of pravastatin 10 mg or simvastatin 5 mg once daily (20 mg pravastatin and 10 mg simvastatin in severe hypercholesterolemia). Primary endpoints were major coronary events: sudden cardiac death, fatal and nonfatal AMI, and nonfatal events such as development of unstable angina, CABG, angioplasty, and/or coronary stenting. Patients were randomized to treatment with EPA (EPA+ statin) or statin alone (controls).

After a mean follow-up of 4.6 years, the primary endpoint had occurred in 262 patients (2.8%) in the EPA group and 324 patients (3.5%) in the control group, with a

Trial	Population	N	Follow-up, y	Interventions	Baseline Statin Use, %	Baseline LDL-C, mg/dL	Baseline TG, mg/dL	Primary Endpoint	Risk Ratio (95% CI)	Adverse Events[a]
Diet and/or supplement										
DART	Recent MI (men)	2033	2	Fatty fish 2 servings/week or supplement EPA + DHA 855 mg/d; total fat 30% of calories, PUFA/SFA ratio 1.0; cereal fiber 18 g/d; no diet advice	NR	TC 250	NR	Total mortality CHD mortality	RR0.71 (0.54–0.93) RR 0.84 (0.66–1.07)	NR
EPA + DHA (low dose)										
GISSI-P	Recent MI	11,324	3.5	EPA + DHA 850 mg/d; vitamin E 300 mg/d; n-3 + vitamin E; no supplement	Cholesterol-lowering agents: 5 (end of study: 46)	137	162	Death or nonfatal MI or nonfatal stroke	RR 0.90 (0.82–0.99) [2-way analysis]; RR 0.85 (0.74–0.98) [4-way analysis]	GI disturbance, nausea
GISSI-HF	Chronic HF	6975	3.9	EPA + DHA 850 mg; placebo (unspecified)	23	TC 188	126	Death Death or CV hosp	HR 0.91 (0.833–0.998) HR 0.92 (0.849–0.999)	GI disturbance
OMEGA	Recent MI	3851	1	DHA + EPA 840 mg; placebo	94–95	NR	NR	SCD	&$$$;DR 0.95 (0.56–1.60)	Neoplasms (19 vs. 8), cardiac device therapeutic procedures (16 vs. 2)
Alpha-omega-3	Prior MI	4837	3.3	400 mg EPA + DHA ± 2 g ALA; 2 g ALA; placebo	Baseline lipid-lowering agents: 85–87	99–102	144–150	Fatal/nonfatal CV events or revasc	HR: 1.01 (0.87–1.17)	GI symptoms, prostate cancer incidence, death

Trial	Population	N	Follow-up, y	Interventions	Baseline Statin Use, %	Baseline LDL-C, mg/dL	Baseline TG, mg/dL	Primary Endpoint	Risk Ratio (95% CI)	Adverse Events[a]
SU.FOL.OM3	Recent acute coronary or cerebral ischemic event	2501	4.7	EPA + DHA 600 mg; placebo + vitamin B	Baseline lipid-lowering agents: 83–87	101–104	108 (no coronary event); 125 (coronary event)	MACE	HR 1.08 (0.79–1.47)	GI disturbance, nausea, cutaneous reactions
ORIGIN	Dysglycemia or high-risk CVD	12,536	6.2	EPA + DHA 840 mg; placebo (olive oil)	53–54	112	140–142	CV death	HR 0.98 (0.87–1.10)	NR
ORIGINALE	ORIGIN participants	4771	8.9	EPA + DHA 840 mg; placebo (olive oil)	57–59	108–109	142	CV death	HR 0.98 (0.88–1.09)	NR
Risk & Prevention	High-risk CVD	12,513	5	EPA + DHA 850 mg; placebo	41	132	150	CV death or CV hosp	HR 0.97 (0.88–1.08)	GI symptoms
ASCEND	Diabetes without CVD	15,480	7.4	DHA + EPA 840 mg; placebo (olive oil)	75	TC 161	NR	Nonfatal MI, nonfatal stroke, TLA, or vascular death	RR 0.97 (0.87–1.08)	Similar between groups
VITAL	Healthy, no history of CVD	25,871	5.3	DHA + EPA 840 mg; placebo	Cholesterol-lowering agents: 37	NR	NR	MI, stroke, or CV death	HR 0.97 (0.85–1.12)	GI symptoms
EPA (hiah& $$$; dose)										
JELIS	TC >250 mg/dL	18,645	4.6	EPA 1800 mg + statin; statin only	97	182	151	Sudden cardiac death, fatal or nonfatal MI, unstable angina, or revasc	HR 0.81 (0.69–0.95)	GI, skin, hemorrhage (cerebral, fundal, epistaxis, subcutaneous) (0.6% vs. 1.1%, $P = 0.0006$)

Trial	Population	N	Follow-up, y	Interventions	Baseline Statin Use, %	Baseline LDL-C, mg/dL	Baseline TG, mg/dL	Primary Endpoint	Risk Ratio (95% CI)	Adverse Events[a]
REDUCE-IT	CVD or diabetes + additional CVD risk	8179	4.9	EPA 4000 mg: placebo (mineral oil)	100	75	216	CV death, nonfatal MI, nonfatal stroke, revasc, or unstable angina	HR 0.75 (0.68–0.83)	Hosp for atrial fibrillation and atrial flutter (3.1% vs. 2.1%, $P = 0.004$), serious bleeding events (2.7% vs. 2.1%, $P = 0.06$)

Modified from Wu H et al. [7]. A Not significantly different between treatment groups except as noted. ALA, alpha-linolenic acid; ASCEND, A Study of Cardiovascular Events in Diabetes; CHD, coronary heart disease; CI, confidence interval; CV, cardiovascular; CVD, cardiovascular disease; DART, Diet and Reinfarction Trial; DHA, docosahexaenoic acid; EPA, eicosapentaenoic acid; GI, gastrointestinal; GISSI-P, Gruppo Italiano per lo Studio della Sopravvivenza nell'Infarto Miocardico Prevenzione; GISSI-HF, GISSI–Heart Failure; HF, heart failure; hosp, hospitalization; HR, hazard ratio; JELIS, Japan EPA Lipid Intervention Study; LDL-C, low-density lipoprotein cholesterol; MACE, major adverse cardiac events; MI, myocardial infarction; NR, not reported; OR, odds ratio; ORIGIN, Outcome Reduction with an Initial Glargine Intervention; ORIGINALE, Outcome Reduction with an Initial Glargine Intervention Legacy Effects; PUFA, polyunsaturated fatty acid; REDUCE-IT, Reduction of Cardiovascular Events with Icosapent Ethyl–Intervention Trial; revasc, revascularization; RR, relative risk; SCD, sudden coronary death; SFA, saturated fatty acid; SU.FOL.OM3, Supplementation with Folate, Vitamin B6 and B12 and/or Omega-3 Fatty Acids trial; TC: total cholesterol; TG, triglyceride; TIA, tran-sient ischemic attack; VITAL, Vitamin D and Omega-3 Trial; y, years.

Table 3.
Reviewed outcomes trials of n-3 PUFAs.

relative 19% reduction in major coronary events (p = 0.011). The 25% reduction in C-LDL in both groups after treatment was not found to be a significant factor in reducing the risk of events. The risk of unstable angina and nonfatal coronary events was reduced by 19% in the EPA group, while sudden cardiac death and death from coronary causes were overlapping in the 2 groups.

In patients with coronary artery disease (secondary prevention), EPA reduced major coronary events by 19% compared with controls (8.7% vs. 10.7%; p = 0.048). In patients free of coronary artery disease, EPA did not significantly reduce major coronary events (−18%; 1.4% vs. 1.7% of controls; p = 0.132). The study results suggested that EPA was a promising treatment for primary and secondary prevention of major coronary events, especially nonfatal coronary events (**Figure** 7).

A sub-analysis of JELIS in 2012, showing 38% (p = 0.007) reduction in coronary events in patients who did not reach target values of C-LDL and non-C-HDL, concluded that EPA supplementation was also more effective in patients who did not reach target values of cholesterolemia.

REDUCE-IT (2019). Multicenter, randomized, placebo-controlled, double-blind study [34] enrolled and followed, on average for 4.9 years, 8179 patients diagnosed with CV disease (70.7%; secondary prevention) or with diabetes mellitus and other risk factors on statin therapy. Patients had to have triglyceridemia between 135 and 499 mg/dL and C-LDL between 41 and 100 mg/dL at baseline.

The aim of the study was to evaluate the potential benefits of EPA ethyl ester on CV outcomes in patients with elevated triglyceridemia levels. Patients were randomized to receive 2000 mg of the EPA ethyl ester twice daily (with a total dose of 4 g) or placebo (mineral oil).

The primary composite endpoint was the association of CV death, nonfatal AMI, nonfatal stroke, coronary revascularization, or unstable angina. Key secondary endpoints were the association of CV death, nonfatal AMI, and nonfatal stroke.

The primary endpoint occurred in 17.2% of patients in the EPA group compared with 22% of controls (HR 0.75; 95% CI, 0.68–0.83; p < 0.001). Secondary endpoints were observed in 11.2% of EPA and 14.8% of placebo patients, respectively (HR 0.74; 95% CI, 0.65–0.83; p < 0.001). The rate of ischemic events was significantly lower in the EPA group than in the placebo group, as was the CV mortality rate (4.3% vs. 5.2%; HR 0.80; 95% CI, 0.66–0.98; p = 0.03).

Regarding the safety profile of EPA, it was observed that hospitalization for atrial fibrillation or flutter occurred in a significant proportion of EPA-treated patients

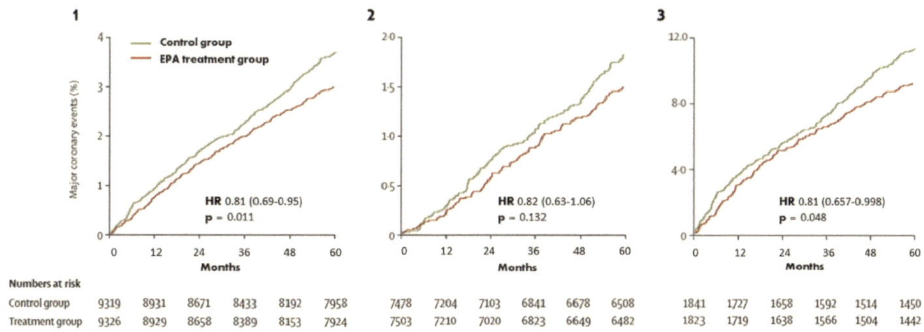

Figure 7.
Kaplan-Meier estimates of incidence of coronary events in the total study population (panel 1), the primary prevention arm (panel 2), and the secondary prevention arm (panel 3). Modified from Yokoyama et al. [33]. HR = Hazard ratio.

(3.1% vs. 2.1%; p = 0.004). Serious bleeding events occurred in 2.7% of EPA patients and 2.1% of controls (p = 0.06).

The study shows that in patients with elevated triglyceridemia, even with statin therapy, the risk of ischemic events and CV death is further reduced by EPA supplementation (**Figure 8**).

STRENGTH (2020). Multicenter, randomized, double-blind study [35], which evaluated the effects of carboxyl formulation (CA) of EPA and DHA on CV outcomes in adult patients on statin therapy for at least 4 weeks and with elevated CV risk, defined by the presence of:

1. Coronary, carotid, aortic, or peripheral atherosclerotic CVM (secondary prevention);

2. Diabetes mellitus type I or type II (age > 40 years for men and > 50 years for women) and at least one additional risk factor: smoking, hypertension, hsPCR >2 mg/L, or moderate albuminuria;

3. High-risk patients (age > 50 years for men and > 60 years for women) with at least one risk factor, including: family history of CAD, chronic smoking, hsPCR >2 mg/L, impaired renal function, or a coronary calcium score > 300 Agatston units (primary prevention).

A total of 13,078 patients were enrolled in 675 community and teaching hospitals and 22 states in North America, Europe, South America, Asia, Australia, New Zealand, and South Africa. All included patients had high TRI values (between 180 and 499 mg/dL) and low C-HDL values (<42 mg/dL in men, <47 mg/dL in women). More than 50% of the enrolled patients met the criteria for secondary CV prevention. Patients were randomized to receive 4 grams/day of CA n-3 PUFAs (n = 6539) or corn oil placebo (n = 6539), in combination with usual medical therapy.

The primary composite efficacy endpoint was an association of CV death, nonfatal AMI, nonfatal stroke, coronary revascularization, and hospitalization for unstable angina. Secondary endpoints included (1) the association of CV death, nonfatal IMA, nonfatal stroke, coronary revascularization, and hospitalization for unstable angina in

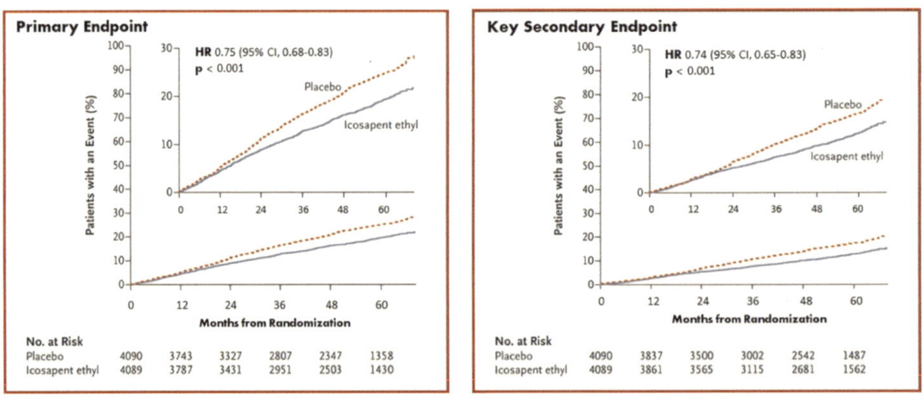

Figure 8.
Cumulative incidence of cardiovascular events in REDUCE-IT trial. Modified from Bhatt et al. [34]. HR = Hazard ratio.

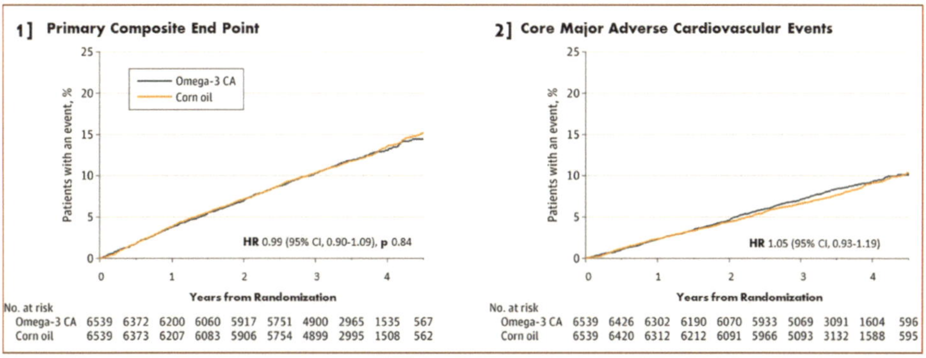

Figure 9.
Time to first incidence of any component of the primary composite end point (1) and time to Core major adverse cardiovascular events (2). Modified from Nicholls SJ et al. [35]. 1, the primary composite end point consisted of cardiovascular death, nonfatal myocardial infarction, nonfatal stroke, coronary revascularization and hospitalization for unstable angina. 2, Core major adverse cardiovascular events included cardiovascular death, nonfatal myocardial infarction and nonfatal stroke.

patients with pre-existing CVD at baseline, (2) the association of CV death, nonfatal IMA, and nonfatal stroke in the whole cohort of patients and in patients with pre-existing CVD at baseline, (3) the association of cardiac death, nonfatal AMI, coronary revascularization, and hospitalization for unstable angina in the entire cohort of patients and in those with pre-existing CVD at baseline, (4) CV death in the entire cohort of patients and in those with pre-existing CVD at baseline, and (5) all-cause mortality in all patients and in those with pre-existing CVD at baseline.

When 1384 patients developed a primary endpoint event (compared with 1600 expected), the trial was stopped early on the basis of a biased analysis of the data, which showed a low likelihood of clinical benefit of n-3 PUFA CA compared with placebo. Study population baseline characteristics were as follows: mean age 62.5 + 9.0 years; 35% women; 70% with diabetes; median C-LDL 75.0 mg/dL; median triglyceride levels 240 mg/dL; median HDL-C levels 36 mg/dL; and median hsPCR 2.1 mg/L. Out of 12,633 patients (96.6%) who completed the trial, the primary end point occurred in 785 (12.0%) treated with n-3 PUFAs compared with 795 (12.2%) treated with corn oil (HR, 0.99 [95% CI, 0.90–1.09]; P = 0.84). There was, in addition, a high rate of gastrointestinal adverse events in the treatment group (24.7% vs. 14.7%).

Therefore, the results of this study do not support the use of PUFA n-3 formulations for improving clinical outcome in patients at high CV risk and on statin therapy (**Figure 9**).

5.1 REDUCE-IT and STRENGTH trials: Why conflicting results?

Several hypotheses have been developed to interpret the reasons that led REDUCE-IT and STRENGTH (two high-quality studies that employed full doses of n-3 PUFAs), to deeply conflicting results [36].

1. One possible explanation theorizes that in the REDUCE-IT study, EPA did not reduce the risk of CV events but, rather, its comparator (mineral oil) increased them. An increase in C-LDL, apoB, and hsPCR associated with mineral oil seems to support this interpretation. However, a U.S. FDA review of the REDUCE-IT

results concluded that this hypothesis could explain only part of the highlighted difference between EPA and mineral oil.

2. A second consideration concerns the use in STRENGTH of CA's formulation of n-3 PUFAs composed of both EPA and DHA, in contrast to studies conducted with purified EPA. Although EPA levels in plasma and red blood cells were higher in REDUCE-IT than in STRENGTH, it is unclear whether these differences are sufficient to explain the observed results. This doubt is accentuated by the fact that in STRENGTH trial there was no significant reduction in the risk of CV events in patients with higher levels of EPA, compared with those with lower levels. Furthermore, the reduction in TRI after 12 months was 18 percent in both trials, suggesting a similar effect of the formulations used.

3. Another food for thought is the fact that STRENGTH was discontinued during the first phase of the COVID-19 pandemic and that the "end-of-treatment" visits were conducted *via* telephone contact in order to allow the trial to be terminated as early and orderly as possible. This may have compromised the integrity of the study itself.

To resolve the discrepancies between STRENGTH and REDUCE-IT, regulatory agencies such as the FDA could authorize a post-marketing clinical trial comparing high-dose EPA vs. corn oil in patients at high risk of developing CV events.

6. Real word data about n-3 PUFAs

TG-REAL (2020). Longitudinal, retrospective cohort study that used 3 administrative databases from 3 Italian local health authorities [37]. It included 158,042 individuals with at least one serum TG assay between January 1, 2010 and December 31, 2015. Individuals with normal TG values (<150 mg/dL; n = 142,289) were compared with patients with high values (150–500 mg/dL; n = 15,558) and those with very high values (>500 mg/dL; n = 195).

The outcomes were atherosclerotic CV events (ASCVD) and all-cause mortality. Overall, the incidences of ASCVD and all-cause mortality were 7.2 and 17.1 per 1000 individual-years, respectively. After multivariate correction for potential confounding factors, in individuals with high and very high TG values, there was a significant increased risk of all-cause mortality (HR = 1.49 (95 percent CI 1.36–1.63), $p < 0.001$, and HR = 3.08 (95 percent CI 1.46–6.50), $p < 0.01$, respectively), and of ASCVD (HR = 1.61 (95 percent CI 1.43–1.82), $p < 0.001$, and HR = 2.30 (95 percent CI 1.02–5.18), $p < 0.05$, respectively).

The TG-REAL study captured real-world data related to a large cohort of patients with low-to-moderate CV risk, demonstrating how moderate to severely elevated TG values are associated with a significant increased risk of ASCVD and all cause death.

7. Final considerations

Following the results of the REDUCE-IT trial, the regulatory agencies FDA and EMA have given their approval for the use of n-3 PUFAs in patients at high CV risk in order to reduce atherosclerotic-based events. The recommendations for the use of n-3 PUFAs in the more recent European and US guidelines are shown in **Table 4** [9].

An Update is expected from the European Scientific Societies. In fact, in the 2019 European Society of Cardiology (ESC) guidelines, the recommendation for the use of n-3 PUFAs in high-risk patients with hypertriglyceridemia is still class IIa, despite the positive results that have recently emerged from clinical trials.

Medical society	Date	Guidelines/standards/advisory statements
American Diabetes Association – ADA	March 2019	ADA's Standards of Medical Care updated Section 10, Treatment of Other Lipoprotein Fractions or Targets, states:
		In patients with ASCVD or other cardiac risk factors on a statin with controlled LDL-C but elevated TG (135–499 mg/dL), the addition of icosapent ethyl should be considered to reduce CV risk—Level A
		It should be noted that data are lacking with other omega-3 fatty acids, and the results of REDUCE-IT should not be extrapolated to other products
		Combination therapy (statin/fibrate) has not been shown to improve atherosclerotic CV disease and is generally not recommended—Level A
		Combination therapy (statin/niacin) has not been shown to provide additional CV benefit above statin therapy alone, may increase the risk of stroke with additional side effects, and is generally not recommended—Level A
American Heart Association – AHA Science Advisory	August 2019	The AHA Issued a Scientific Advisory that dietary supplements are not recommended and that positive outcomes results were demonstrated in REDUCE-IT:
		The use of omega-3 fatty acids (4 g/day) for improving ASCVD risk in patients with hypertriglyceridemia is supported by a 25% reduction in MACE in REDUCE-IT
		The potency, quality, and efficacy of dietary supplement are not initially reviewed or approved, nor are they subsequently monitored or assured by the FDA; thus, they are not indicated for the treatment of disease
		In the treatment of patients with very high TG levels with 4 g/day, EPA + DHA agents reduce TG by ≥30% with concurrent increases in LDL-C, whereas EPA-only does not raise LDL-C in patients with very high TG levels
European Atherosclerosis Society/ European Society of Cardiology – EAS/ESC	August 2019	The EAS and ESC jointly updated patient treatment guidelines to state: In high-risk (or above) patients with TG between 1.5 and 1.6 mmol/L (135–499 mg/dL) despite statin treatment, n-3 PUFAs (Icosapent ethyl 2 × 2 g/day) should be considered in combination with statins – Class IIa, Level B
National Lipid Association – NLA	September 2019	NLA Position on the Use of Icosapent Ethyl in High- and Very High-Risk Patients states:

Medical society	Date	Guidelines/standards/advisory statements
		For patients, ≥45 years with clinical ASCVD, or ≥ 50 years with type 2 diabetes requiring medication and ≥ 1 additional risk factor, and fasting TGs 135–499 mg/dL on a maximally tolerated statin, with or without ezetimibe, treatment with icosapent ethyl is recommended for ASCVD reduction – Class 1, Level B-R
American Heart Association Scientific Statement	April 2020	The AHA recommends consideration of icosapent ethyl in patients with type 2 diabetes for further CV risk reduction when TGs remain elevated despite maximally tolerated statin for management of coronary artery disease
New FDA-approved Indication:		
FDA-approved label expansion beyond TG lowering	December 2019	Icosapent ethyl is indicated as an adjunct to **maximally tolerated statin therapy** to reduce the risk of MI, stroke, coronary revascularization, and unstable angina requiring hospitalization in adult patients with elevated TG levels (≥ **150 mg/dL**) and established CV disease or diabetes mellitus and **2 or more additional risk factors** for CV disease[a]

[a]Bold text highlights aspects of expanded label indication that notably differ from language in updated guidelines and REDUCE-IT entry criteria.
Modified from Trivedi K et al. [11]. ADA, American Diabetes Association; AHA, American Heart Association; CV, cardiovascular; EAS, European Atherosclerosis Society; EPA, eicosapentaenoic acid; ESC, European Society of Cardiology; FDA, Food and Drug Administration; LDL-C, low-density lipoprotein cholesterol; MI, myocardial infarction; NLA, National Lipid Association; REDUCE-IT, Reduction of Cardiovascular Events with Icosapent Ethyl–Intervention Trial; TG, triglyceride.

Table 4.
Statements from medical societies and the US FDA regarding IPE (pure EPA) post-REDUCE-IT.

Currently, the strategy for the management and prevention of CV diseases follows the so-called "ABCDEF" scheme.

In that strategy:

- **A** corresponds to CV risk assessment (Assessment of risk), the need to set antiplatelet therapy (Antiplatelet therapy), and the need to manage atrial fibrillation (Atrial fibrillation management).
- **B** corresponds to the management of Blood pressure.
- **C** corresponds to Cholesterol management and cessation of cigarette smoking.
- **D** corresponds to Diet and lifestyle modification, as well as diabetes management.
- **E** corresponds to Exercise.
- **F** to the management of heart Failure.

Based on current evidence, it is believed that under "E," along with exercise, the intake of EPA and IPE, rather than fish oil dietary supplements, should also be considered as a "new entry."

It is also important to note that previous studies on n-3 PUFAs, in which there was a goal of lowering TG levels, have not obtained concordant results regarding their efficacy in reducing CV events. These conflicting results seem attributable to several factors, including frequent use of inappropriate drug dosages, use of different formulations, and/or inadequate selection of the target population. Therefore, it is of great relevance to find answers to the outstanding questions and to overcome the biases highlighted in past clinical trials. Such doubts or questions could only be clarified by performing comparison studies between different available formulations, such as those used in the REDUCE-IT and STRENGTH clinical trials, in order to fully understand the real differences in terms of efficacy and safety of these drugs.

Finally, it is important to point out that IPE therapy has been shown to be closely related to a reduction in the residual risk of total ischemic events in patients with ASCVD or with type II diabetes mellitus and other CV risk factors, already on statin therapy. These evidences would therefore support a preferential indication for the use of IPE over currently marketed formulations.

Conflict of interest

The authors declare no conflict of interest.

Author details

Savina Nodari* and Francesco Fioretti
Cardiology Unit, Department of Medical and Surgical Specialties, Radiological Sciences and Public Health, University and Spedali Civili Hospital of Brescia, Brescia, Italy

*Address all correspondence to: savina.nodari@unibs.it

References

[1] Wu H, Xu L, Ballantyne CM. Dietary and pharmacological fatty acids and cardiovascular health. The Journal of Clinical Endocrinology and Metabolism. 2020;**105**(4):1030-1045. DOI: 10.1210/clinem/dgz174

[2] Mach F, Baigent C, Catapano AL, Koskinas KC, Casula M, Badimon L, et al. ESC/EAS guidelines for the management of dyslipidaemias: Lipid modification to reduce cardiovascular risk. Eur Heart J. 2019, 2020;**41**(1):111-188. DOI: 10.1093/eurheartj/ehz455. Erratum in: Eur Heart J. 2020;41(44): 4255

[3] IUPAC-IUB Commission on Biochemical Nomenclature (CBN). The nomenclature of lipids. Recommendations, 1976. European Journal of Biochemistry. 1977;**79**:11-21

[4] Tvrzicka E, Kremmyda LS, Stankova B, Zak A. Fatty acids as biocompounds: Their role in human metabolism, health and disease–a review. Part 1: Classification, dietary sources and biological functions. Biomedical Papers of the Medical Faculty of the University Palacky, Olomouc, Czech Republic. 2011;**155**(2):117-130. DOI: 10.5507/bp.2011.038

[5] Mozaffarian D. Does alpha-linolenic acid intake reduce the risk of coronary heart disease? A review of the evidence. Alternative Therapies in Health and Medicine. 2005;**11**(3):24-30

[6] Balk E, Adams G, Langberg V, Halladay C, Chung M, Lin L, et al. Omega-3 fatty acids and cardiovascular disease: An updated systematic review. Evid Rep Technol Assess (Full Rep). 2016;**223**:1-1252

[7] Del Gobbo LC, Imamura F, Aslibekyan S, Marklund M, Virtanen JK, Wennberg M, et al. Cohorts for heart and aging research in genomic epidemiology (CHARGE) fatty acids and outcomes research consortium (FORCe). ω-3 polyunsaturated fatty acid biomarkers and coronary heart disease: Pooling project of 19 cohort studies. JAMA Intern Med. 2016;**176**:1155-1166

[8] Marklund M, Wu JHY, Imamura F, Del Gobbo LC, Fretts A, de Goede J, et al. Biomarkers of dietary omega-6 fatty acids and incident cardiovascular disease and mortality: An individual-level pooled analysis of 30 cohort studies. Circulation. 2019;**139**:2422-2436

[9] Trivedi K, Le V, Nelson JR. The case for adding eicosapentaenoic acid (icosapent ethyl) to the ABCs of cardiovascular disease prevention. Postgraduate Medicine. 2021;**133**(1):28-41. DOI: 10.1080/00325481.2020.1783937

[10] Wei Y, Meng Y, Li N, Wang Q, Chen L. The effects of low-ratio n-6/n-3 PUFA on biomarkers of inflammation: A systematic review and meta-analysis. Food & Function. 2021;**12**(1):30-40. DOI: 10.1039/d0fo01976c

[11] Russo GL. Dietary n-6 and n-3 polyunsaturated fatty acids: From biochemistry to clinical implications in cardiovascular prevention. Biochemical Pharmacology. 2009;**77**(6):937-946. DOI: 10.1016/j.bcp.2008.10.020

[12] Mozaffarian D, Wu JH. Omega-3 fatty acids and cardiovascular disease: Effects on risk factors, molecular pathways, and clinical events. Journal of the American College of Cardiology. 2011;**58**(20):2047-2067. DOI: 10.1016/j.jacc.2011.06.063

[13] Bays HE, Tighe AP, Sadovsky R, Davidson MH. Prescription omega-3

fatty acids and their lipid effects: Physiologic mechanisms of action and clinical implications. Expert Review of Cardiovascular Therapy. 2008;**6**:391-409

[14] Harris WS, Bulchandani D. Why do omega-3 fatty acids lower serum triglycerides? Current Opinion in Lipidology. 2006;**17**:387-393

[15] Backes J, Anzalone D, Hilleman D, Catini J. The clinical relevance of omega-3 fatty acids in the management of hypertriglyceridemia. Lipids in Health and Disease. 2016;**15**(1):118. DOI: 10.1186/s12944-016-0286-4

[16] Wei MY, Jacobson TA. Effects of eicosapentaenoic acid versus docosahexaenoic acid on serum lipids: A systematic review and meta-analysis. Current Atherosclerosis Reports. 2011;**13**:474-483

[17] Bays HE, Ballantyne CM, Kastelein JJ, Isaacsohn JL, Braeckman RA, Soni PN. Eicosapentaenoic acid ethyl ester (AMR101) therapy in patients with very high triglyceride levels (from the multi-center, plAcebo-controlled, randomized, double-blINd, 12-week study with an open-label extension [MARINE] trial). The American Journal of Cardiology. 2011;**108**:682-690

[18] Bays HE, Ballantyne CM, Braeckman RA, Stirtan WG, Soni PN. Icosapent ethyl, a pure ethyl ester of eicosapentaenoic acid: Effects on circulating markers of inflammation from the MARINE and ANCHOR studies. American Journal of Cardiovascular Drugs. 2013;**13**:37-46

[19] Lee MW, Park JK, Hong JW, Kim KJ, Shin DY, Ahn CW, et al. Beneficial effects of omega-3 fatty acids on low density lipoprotein particle size in patients with type 2 diabetes already under statin therapy. Diabetes and Metabolism Journal. 2013;**37**:207-211

[20] Ballantyne CM, Bays HE, Kastelein JJ, Stein E, Isaacsohn JL, Braeckman RA, et al. Efficacy and safety of eicosapentaenoic acid ethyl ester (AMR101) therapy in statin-treated patients with persistent high triglycerides (from the ANCHOR study). The American Journal of Cardiology. 2012;**110**:984-992

[21] Nishio R, Shinke T, Otake H, Nakagawa M, Nagoshi R, Inoue T, et al. Stabilizing effect of combined eicosapentaenoic acid and statin therapy on coronary thin-cap fibroatheroma. Atherosclerosis. 2014;**234**(1):114-119. DOI: 10.1016/j.atherosclerosis.2014.02.025

[22] Kashiyama T, Ueda Y, Nemoto T, Wada M, Masumura Y, Matsuo K, et al. Relationship between coronary plaque vulnerability and serum n-3/n-6 polyunsaturated fatty acid ratio. Circulation Journal. 2011;**75**(10):2432-2438. DOI: 10.1253/circj.cj-11-0352

[23] Budoff MJ, Bhatt DL, Kinninger A, Lakshmanan S, Muhlestein JB, Le VT, et al. Effect of icosapent ethyl on progression of coronary atherosclerosis in patients with elevated triglycerides on statin therapy: Final results of the EVAPORATE trial. European Heart Journal. 2020;**41**(40):3925-3932. DOI: 10.1093/eurheartj/ehaa652

[24] Jacobson TA, Ito MK, Maki KC, Orringer CE, Bays HE, Jones PH, et al. National Lipid Association recommendations for patient-centered management of dyslipidemia: Part 1 - executive summary. Journal of Clinical Lipidology. 2014;**8**:473-488

[25] Balk EM, Lichtenstein AH, Chung M, Kupelnick B, Chew P, Lau J. Effects

of omega-3 fatty acids on serum markers of cardiovascular disease risk: A systematic review. Atherosclerosis. 2006;**189**:19-30

[26] Bays HE, Braeckman RA, Ballantyne CM, Kastelein JJ, Otvos JD, Stirtan WG, et al. Icosapent ethyl, a pure EPA omega-3 fatty acid: Effects on lipoprotein particle concentration and size in patients with very high triglyceride levels (the MARINE study). Journal of Clinical Lipidology. 2012;**6**:565-572

[27] Ballantyne CM, Braeckman RA, Bays HE, Kastelein JJ, Otvos JD, Stirtan WG, et al. Effects of icosapent ethyl on lipoprotein particle concentration and size in statin-treated patients with persistent high triglycerides (the ANCHOR study). Journal of Clinical Lipidology. 2015;**9**:377-383

[28] Mason RP, Libby P, Bhatt DL. Emerging mechanisms of cardiovascular protection for the Omega-3 fatty acid Eicosapentaenoic acid. Arteriosclerosis, Thrombosis, and Vascular Biology. 2020; **40**(5):1135-1147. DOI: 10.1161/ATVBAHA.119.313286

[29] Zargar A, Ito MK. Long chain omega-3 dietary supplements: A review of the National Library of medicine herbal supplement database. Metabolic Syndrome and Related Disorders. 2011;**9**: 255-271

[30] Nodari S, Triggiani M, Campia U, Manerba A, Milesi G, Cesana BM, et al. n-3 polyunsaturated fatty acids in the prevention of atrial fibrillation recurrences after electrical cardiovesrion: A prospective, randomized study. Circulation. 2011;**124**(10):1100-1106. DOI: 10.1161/CIRCULATIONHA.111.022194

[31] Nodari S, Triggiani M, Campia U, Dei CL. Omega-3 Polyunsaturated Fatty Acid Supplementation: Mechanism and Current Evidence in Atrial Fibrillation. Journal of Atrial Fibrillation. 2012;**5**(4): 718. DOI: 10.4022/jafib.718

[32] Gruppo Italiano per lo Studio della Sopravvivenza nell'Infarto miocardico. Dietary supplementation with n-3 polyunsaturated fatty acids and vitamin E after myocardial infarction: Results of the GISSI-Prevenzione trial. Lancet. 1999;**354**(9177):447-455 Erratum in: Lancet 2001 Feb 24;357(9256):642. Erratum in: Lancet. 2007 Jan 13;369 (9556):106

[33] Yokoyama M, Origasa H, Matsuzaki M, Matsuzawa Y, Saito Y, Ishikawa Y, et al. Effects of eicosapentaenoic acid on major coronary events in hypercholesterolaemic patients (JELIS): A randomised open-label, blinded endpoint analysis. Lancet. 2007;**369** (9567):1090-1098. DOI: 10.1016/S0140-6736(07)60527-3 Erratum in: Lancet. 2007 Jul 21;370(9583):220

[34] Bhatt DL, Steg PG, Miller M, Brinton EA, Jacobson TA, Ketchum SB, et al. Cardiovascular risk reduction with Icosapent ethyl for hypertriglyceridemia. The New England Journal of Medicine. 2019;**380**(1):11-22. DOI: 10.1056/NEJMoa1812792

[35] Nicholls SJ, Lincoff AM, Garcia M, et al. Effect of high-dose Omega-3 fatty acids vs corn oil on major adverse cardiovascular events in patients at high cardiovascular risk: The STRENGTH randomized clinical trial. Journal of the American Medical Association. 2020;**324** (22):2268-2280. DOI: 10.1001/jama.2020.22258

[36] Curfman G. Do Omega-3 fatty acids benefit health? Journal of the American Medical Association. 2020; **324**(22):2280-2281. DOI: 10.1001/jama.2020.22898

[37] Arca M, Veronesi C, Derasmo L, Borghi C, Colivicchi F, De Ferrari GM, et al. Association of Hypertriglyceridemia with all-cause mortality and atherosclerotic cardiovascular events in a low-risk Italian population: The TG-REAL retrospective cohort analysis. Journal of the American Heart Association. 2020;**9**(19):e015801. DOI: 10.1161/JAHA.119.015801

Chapter 4

Omega-3 Fatty Acids: Novel Insight into Cardiovascular Events, Cardiovascular Disease (CVD), and Cardiac Arrhythmias

Muralidharan Velappan and Deecaraman Munusamy

Abstract

It is a common knowledge that fish is a significant source of docosahexaenoic acid and eicosapentaenoic acid, two long-chain omega-3 fatty acids that have been linked to improve cardiovascular health in general. The cardiac function of humans is benefited by omega-3 fatty acids found in fish eating. Previous studies have shown that eating fish in moderation lowers the risk of coronary heart disease. Recent epidemiological research on the relationship between fish consuming and coronary disease have produced mixed results. Omega-3 fatty acids may not, according to a recent study, lower the incidence of cardiovascular events, strokes, cardiac arrhythmias, or fatalities from coronary heart disease; consequently, it continues to be a contentious issue.

Keywords: fish consumption, omega-3 fatty acids, epidemiology, cardiovascular disease, deaths

1. Introduction

Most of the types of fats that the human body requires can be produced by it from other fats or raw materials. The case with omega-3 fatty acids is different; these are necessary fats that the body cannot produce on its own and must obtain from food. Fish, vegetable oils, almonds, walnuts, flaxseeds, flaxseed oil, and green vegetables are among the foods high in omega-3 fatty acids. Additionally, they have a significant role in human physiology [1]. According to a number of study reports, eating seafood at least once a week lowers your risk of heart disease compared to eating it occasionally or never [2]. In 2017, globally, cardiovascular disease (CVD) was the top disease that leads to premature death and contributes to 8.9 million deaths from coronary heart disease (CHD), which is a major cause of death in the UK and worldwide. CHD is sometimes called ischaemic heart disease or coronary heart disease.

According to past studies, the cardioprotective benefits of eating fish have long attracted scientific attention. In recent years, numerous studies have focused on long-chain omega-3 polyunsaturated fatty acids (PUFAs) found in finfish, such as eicosapentaenoic acid (EPA), docosapentaenoic acid (DPA), 22:5 n-3, and docosahexaenoic acid (DHA), 22:6 n-3 [3]. On the other hand, this meta-analysis suggests that consuming fish twice or three times per week may protect against fatal CHD and sudden cardiac death. Review findings from Refs. [4, 5] claim that the dose response analysis on CHD incidence and mortality were reduced by 4% with a 20 g/day increase in fish consumption.

The tentative contribution from polyunsaturated fatty acids (PUFA) or omega-3 PUFA are the healthy lipid molecules that have two or more carbon-carbon double bonds and it is derived from fish oil. Omega 3 fatty acids are of 3 types: alpha-linolenic acid, eicosapentaenoic acid and docosahexaenoic acid. In recent days, these findings have received much scientific attention, with the word "omega-3" fetching a synonym for fish mostly from salmon and other cold-water fish [6]. According to the author [7] reported, there is conflicting information regarding the impact of fish oil on LDL cholesterol (LDL-C), but he also emphasizes the overwhelming majority of studies that have discovered decreased LDL-C levels. Consequently, regular intake of fish oil shows LDL-C levels either do not change or may increase. N-3 PUFA may also lower CVD through acting as antithrombotic, anti-inflammatory, and antiarrhythmic drugs, as well as by improving endothelial function. According to the research carried out by Kris-Etherton et al. [8] and her co-workers, N-3 PUFA actions and the relation to CVD are schematically presented in **Figure 1**.

Furthermore, the study reports from Refs. [6, 9] evaluated that 116,764 people had a diet of fish or fish oil and CHD mortality. Their review indicated that fish consumption and CHD mortality had an inverse dose-dependent relationship, with an optimal dose of about 40/60 g of fish per day-1, but only in those at high risk for CHD. Several findings indicated that low-risk people who eat fish and maintain healthy lives would not experience any additional protection against CHD. According to past editorial reports [10], regular fish or fish oil diet reduces CVD and overall mortality in patients recovering from myocardial infarction (MI). In addition, Popoff et al. [11] also mentioned that the adult patients with acute myocardial infarction (AMI) are unlikely to notice any notable benefits from omega 3 fatty acid treatment after myocardial infarction. In addition, other meta-analyses demonstrate that CHD death

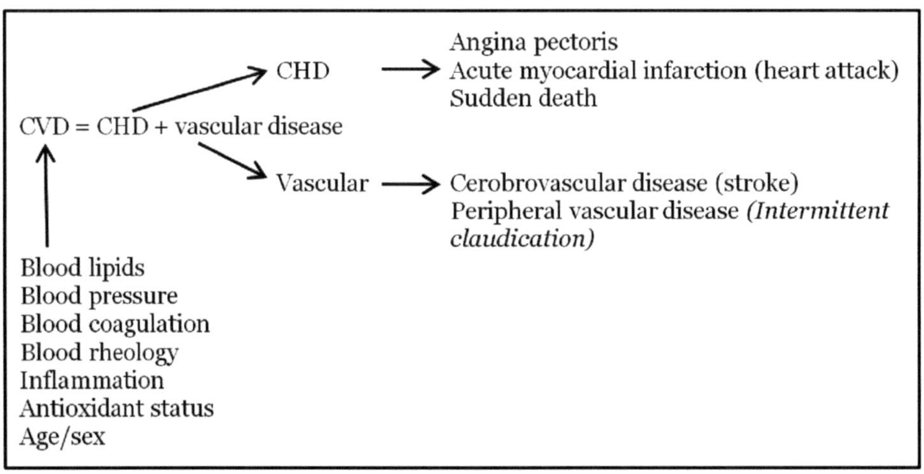

Figure 1.
Source: Fish and cardiovascular health by Undeland et al. [6].

or incidence is not correlated with one another [12–15]. In people who are at high risk for or who already have heart disease, taking omega-3 supplements increases the risk of atrial fibrillation (A-fib), a condition characterised by an irregular and frequently very rapid heartbeat (arrhythmia) that can lead to blood clots in the heart, and there is no correlation between the intake of omega-6 fatty acids and the level of linoleic acid (LA) in circulating lipids and triglycerides. These findings are based on the novel study cited in Ref. [16].

Benefits of n-3 PUFA for coronary heart disease mortality and sudden cardiac death appear to be most consistent in prospective observational studies and adequately powered randomised clinical trials. Less is known about potential impacts on other cardiovascular events, such as non-fatal myocardial infarction, ischemic stroke, atrial fibrillation, recurrent ventricular arrhythmias, and heart failure, which have inconsistent evidence from observational studies and/or randomised trials. The relative significance of various physiological and molecular mechanisms, the precise dose-responses of physiological and clinical effects, the question of whether fish oil provides all the health benefits of fish consumption, and the clinical effects of plant-derived n-3 PUFA are all areas of research that need to be addressed [17]. This review's major goals are to assess (a) fish oil supplements and coronary artery disease, (b) the dosage of omega-3 fatty acids for cardiovascular events, (c) the contribution of fish n-3 fatty acids to CVD and potential risks associated with increased fish consumption, and (d) the proportion of people who take fish oil and experience atrial fibrillation or erratic heartbeats.

2. Omega-3 fatty acid dosage for cardiovascular protection

Since the initial American Heart Association (AHA) Science Advisory was published in 1996, significant new findings regarding the benefits of omega-3 fatty acids on (CVD) have been reported. Fatty fish such as mackerel, halibut, salmon, bluefish, mullet, sablefish, menhaden, anchovy, herring, lake trout, coho, and sardines, which give 1 g or more of omega fatty acids per 100 g (3.5 oz) of fish, are the marine sources with the greatest level of omega-3 fatty acids. Additionally, marine-based sources include tuna, seal, and shellfish [18–22]. The Mayo Clinic Proceedings study and recently released Harvard University study on omega-3 intake and cardiovascular benefits and both agree that an optimal target for daily omega-3 intake is upward of 1000 mg. However, given that the typical daily consumption of EPA and DHA in United States is around 100 mg, additionally, the researchers discovered that the cardiovascular advantages seem to grow with dosage. An additional 1000 mg of (EPA) and (DHA) per day decreased the risk of heart attack by 9.0% and cardiovascular disease by 5.8%, respectively. The trial used daily doses ranging from 400 mg to 5500 mg [23]. According to Dr. Carl J. Lavie, a cardiologist at John Ochsner Heart and Vascular Institute in New Orleans and his colleagues reported, people should consider using omega-3 supplements to help them reach total daily intake of 1000–2000 mg as a relatively low-cost, high-impact way to improve heart health with few associated risks [24]. The AHA also advises consuming at least two servings of fatty fish per week. On the other hand, clinical studies recommend consuming a minimum of 2 g of omega-3 fatty acids per day when triglycerides are raised and a maximum of 4 g per day when coronary heart disease is present. Typically, fish oil supplements have 1000 mg of fish oil, which is equivalent to 300–400 mg of EPA and DHA.

3. A debate on pure N-3 polyunsaturated fatty acids for cardiovascular disease

For more than 50 years, experts have disagreed over whether or not long-chain n-3 polyunsaturated fatty acids (n-3 PUFA) are the effective treatments for cardiovascular disease (CVD). Considering that 8% of Americans consume these chemicals on a daily basis and that the worldwide omega-3 market is worth USD 31.4 billion (KWD 9,460,820,000.00), further data may be useful for public health policy-makers as they discuss the efficacy of the practise. Every nation experiences unacceptably high rates of morbidity and mortality from this condition, especially coronary heart disease (CHD), and despite the adoption of cutting-edge treatments, fatal instances continue. In fact, "fish oil" has consistently rated among the most talked-about topics in the lay press over the past few decades. Due to these variables, it is appropriate to update the research on the potential benefits of these substances for both those with and without cardiac illness [25–27].

4. Biomarkers for omega-3 polyunsaturated fatty acids and coronary heart disease

Omega-3 polyunsaturated fatty acids (PUFAs) have been shown to have a preventative effect on clinical risk factors and coronary heart disease (CHD) risk pathways, according to a review of experimental research and randomised clinical trials [26]. Significant conflicts do, however, still exist. First, randomised clinical trials using fish oil supplements for (CHD) cases have produced inconsistent findings [27, 28]. While taking multiple cardiovascular drugs, the bulk of trials only offered supplements for a brief period to people who already had (CHD) or were at high risk for getting it. In these trials, baseline food consumption was not generally examined, and thus, many participants may have received adequate nourishment from other sources. We do not yet know how generalizable they are or how relevant they are for dietary N-3 PUFA effects on primary (CHD) prevention. In addition, in a study by Kasim et al. [29], a number of earlier observational studies found inverse correlations between N-3 PUFAs from seafood and (CHD) death; however, the effects on non-fatal myocardial infarction (MI) or total (CHD) are less clear. According to this review by Friday et al. [30], the majority of research also used self-reported dietary questionnaires, which can lead to bias or inaccuracies in recollection.

Several studies have assessed the objective levels of circulating or tissue N-3 PUFAs, according to a prior review, although these results may be constrained by publication bias. Prior studies of specific biomarkers were also typically insufficient to investigate any potential changes in effects related to participant-level factors, medication usage, or genetic variation. As a result, questions still surround the impact of N-3 PUFAs on CHD. The majority of investigations have examined combined intakes or biomarker levels of long-chain −3 PUFAs [31, 32]. Individual omega-3 PUFAs, such as eicosapentaenoic acid EPA; 20:5-3 and docosapentaenoic acid DPA; 22:5-3 and docosahexaenoic acid DHA; 22:6-3, may, however, have complementary and shared benefits [33]. Similar to this, the majority of prior studies [34] have concentrated on long-chain omega-3 PUFAs produced from seafood, and the potential CHD effects of plant-derived alpha-linolenic acid ALA; 18:3-3 are significantly less well known.

5. Consumption of fish and dietary omega-3 fatty acids and the risk of type-2 diabetes

An increased risk of death is linked to type 2 diabetes (T2D), which also has expensive medical implications. To create successful preventative methods, it is crucial to identify pertinent risk factors and safeguards. However, information on how omega-3 fatty acids affect insulin sensitivity and glucose metabolism has been scarce and inconsistent [35]. According to Refs. [36, 37] observation revealed the deleterious effects of high doses of fish oils on glycaemic control in individuals with T2D. A greater omega-3 fish oil intervention (6 g omega-3/d) was linked in a randomised control experiment to elevate blood glucose levels and impaired insulin sensitivity in T2D patients. However, in a 6-week trial of (T2D) patients, Bowman et al. [38] found no impact of 4 g pure eicosapentaenoic acid (EPA) and docosahexaenoic acid (DHA) on insulin sensitivity or fasting insulin concentrations. It is yet unknown whether EPA and DHA, especially when consuming very small amounts, have long-term impacts on given the brief periods and greater dosages of omega-3 employed in these experiments. Omega-3 fatty acids may increase the risk of (T2D) in the general population, however this is unknown. Although it is unknown whether findings were the same in men and women, data from the Atherosclerosis Risk in Communities (ARIC) research [39] found no correlation between marine omega-3 fatty acids and T2D risk. Researchers made similar findings from the Kouopio Ischemic Heart Disease Risk factor study [40], although these study only included men and 34 cases of incident diabetes. In contrast, the Iowa Women's Health Study [41] showed a threshold effect, where a little increase in the incidence of (T2D) was only seen in the fifth quintile of marine omega-3 fatty acids. Recent research from three prospective cohorts (the Nurses' Health Study I and II and the Health Professionals Follow-Up Study) suggested that women, but not men, might have a slightly higher risk of (T2D) with higher consumption of long-chain omega-3 and seafood [42]. Wu et al. [43] stated that it is unknown if long-chain omega-3 plays a part in the connection between fish and diabetes. Determining whether consuming small-to-moderate amounts of total and particular omega-3 has a detrimental effect on (T2D) risk is still vital. We therefore sought to prospectively examine in the Women's Health Study the association between dietary omega-3 fatty acids, specifically a-linolenic acid (ALA), fish consumption, and incident diabetes [24]. In addition, Zoler [44] also mentioned that the people who were randomised to consume more long-chain omega-3 fats in the form of fish oils had the same risk of diabetes diagnosis as the control group who did not take more fish oil, despite the fact that over 58,000 participants were enrolled in long-term trials and 4% of those participants developed diabetes. People who take and do not take supplemental fish oils had similar levels of blood glucose, insulin, and glycated haemoglobin, markers of how well our bodies handle carbohydrates (glucose metabolism) and significant indicators of diabetes risk.

Fish oils (long-chain omega-3 fatty acids) had no discernible impact on any of these diabetes risk variables. However, there was some limited evidence that suggested persons may experience worsening glucose metabolism when taking high dosages of fish oils. An example of a fat is omega-3.

The food we eat contains trace levels that are necessary for health. Alpha linolenic acid (ALA), eicosapentaenoic acid (EPA), and docosahexaenoic acid are the three main forms of omega 3 fatty acids (DHA). ALA is typically found in plant-based lipids like those found in nuts and seeds (walnuts and rapeseed are rich sources).

EPA and DHA are referred to as long-chain omega-3 fats and are naturally found in fatty fish, such as salmon and fish oils including cod liver oil [45]. Results from 83 randomised controlled trials involving 121,070 people with and without diabetes and all lasting at least 6 months are combined in the systematic review. In studies, published between the 1960s and 2018, participants included men and women from North America, South America, Europe, Australia, and Asia, some of whom were in good health and others who already had diabetes. The study examined the impact on diabetes and glucose metabolism of increasing long-chain omega-3 fats, ALA, omega-6, and polyunsaturated fatty acids (PUFAs) [46].

6. Atrial fibrillation and the percentage of those consuming fish oil

The most prevalent form of abnormal heart rhythm, atrial fibrillation causes the heart to beat erratically and occasionally too rapid. Brown et al. [47] reported that omega-3 supplements are linked to an increased risk of developing atrial fibrillation in persons with high blood lipid level. That is the conclusion of the study, which was recently published in the European Society of Cardiology's journal, European Heart Journal—Cardiovascular Pharmacotherapy (ESC). The most prevalent heart rhythm issue, atrial fibrillation, has been linked in certain clinical studies to omega-3 fatty acids and an increased risk of the condition. According to Hindricks et al. [48], people with the disease have a five-fold higher chance of suffering a stroke. In the investigation, five randomised controlled trials that looked at the impact of omega-3 fatty acid supplementation on cardiovascular outcomes were included. Participants had excessive triglyceride levels and either had cardiovascular disease in the early stages or were at high risk for it. There were 50,277 patients in all received fish oils or a placebo and were monitored for 2–7 years.

Fish oil dosages ranged from 0.84 g to 4 g daily with an incidence rate ratio of 1.37 (95% confidence interval 1.22–1.54; p 0.001), and the researchers discovered that taking supplements of omega-3 fatty acids was substantially more likely to cause atrial fibrillation than taking a placebo. According to the National Health Interview Survey, 2012, the results are notable, and nearly 19 million adult Americans, or 7.8%, are thought to take fish oil supplements. Recent research suggests neither vitamin D nor the omega fatty acids contained in fish oil prevent the development of atrial fibrillation, according to Christine M. Albert, professor of cardiology and chief of the department of cardiology at the Smidt Heart Institute. Confusion among medical professionals and the public resulted from other clinical trials carried out outside of Cedars-Sinai that suggested patients receiving omega-fatty acid treatment had a higher risk of developing atrial fibrillation.

7. Potential flaws in fish-based research

When reading the present review on the connection between fish and sickness, it should be remembered that it frequently has methodological flaws. Hjartaker [49] examined these flaws, for instance, lean fish and fatty fish are frequently considered to be the same type of meal despite having quite different amounts of energy, saturated fatty acids, monounsaturated fatty acids, and polyunsaturated fatty acids. Additionally, fatty fish have slightly lower quantities of water-soluble substances like calcium but higher levels of fat-soluble vitamins (A, D, and a-tocopherol). Fish from freshwater and saltwater habitats are frequently mixed together. Variations related to time of the year that fish are caught, how they are stored and cooked, how much fish is eaten, and the things that go with fishmeal are frequently ignored or not mentioned. Studies involving vegetarians suggest that effects may be brought on by the consumption of an animal protein source rather than fish [50, 51]. Different definitions of sudden death, residual confounding from reference groups with a less healthy lifestyle, variable endpoints evaluated, various study populations, and potential confounding effects from an increase in haemorrhagic stroke are further causes of epidemiological inconsistency. The majority of the fish studies that are now accessible are conducted on men, and Hu et al. [52] concluded women's studies have not been published in studies for a while.

8. Conclusion

The prevalence of CVD is increasing globally, especially in elderly people. The relationship between fish consumption and the incidence of CVD is uncertain, despite the fact that n-3 fatty acids have effects that may favourably influence CVD risk. Consuming tuna or other baked or broiled fish, but not fried fish, is linked to a decreased prevalence of CVD in older persons. Observational data show a modest, inverse relationship between eating fish and long-chain omega 3 fatty acids and the risk of cerebrovascular accidents. The protective impact of fish consumption on cerebrovascular risk may be mediated by a complicated interplay between varieties of nutrients that are frequently found in fish. However, there was no proof that long chain omega 3 fatty acids evaluated as circulating biomarkers in observational studies or supplements in primary and secondary prevention trials had a similar inverse relationship with cerebrovascular illness. Furthermore, study investigation shows associations between fish intake and risk of CVD, poised to become the leading cardiovascular health burden in coming decades. According to the review and the validation of these results in additional studies, further research into potential mechanisms of benefit and risk, with special emphasis on various fishmeal, is warranted.

Acknowledgements

The author is extremely grateful to my mentor, Prof. Dr. M. Deecaraman, for his ongoing support, encouragement, and helpful criticism. The author also thanks Prof. Dr. Ingrid Undeland of Charlmers University of Technology in Goteborg, Sweden, for granting me permission to copy the figures so I could finish this review article.

Conflict of interest

We declare that we have no conflict of interests.

Abbreviations

CHD	coronary heart disease
CVD	cardiovascular disease

PUFAs	polyunsaturated fatty acids
EPA	eicosapentaenoic acid
DPA	docosapentaenoic acid
LDL-C	low-density lipoprotein-cholesterol
N-3 PUFA	N-3 polyunsaturated fatty acids
MI	myocardial infarction
A-fib	atrial fibrillation
LA	linoleic acid
ALA	alpha-linolenic acid
T2D	type-2 diabetes
ARIC	atherosclerosis risk in communities
AHA	American Heart Foundation
ESC	European Society Cardiology
HDL	high-density lipoprotein

Author details

Muralidharan Velappan[1*] and Deecaraman Munusamy[2]

1 Department of Marine Biotechnology, AMET University, Chennai, Tamil Nadu, India

2 Department of Biotechnology, Dr. M.G.R. Educational and Research Institute, Chennai, Tamil Nadu, India

*Address all correspondence to: muralidharanmicro@gmail.com

References

[1] Omega-3 Fatty Acids: An Essential Contribution. Available from: https://www.hsph.harvard.edu/nutritionsource/what-should-you-eat/fats-and-cholesterol/types-of-fat/omega-3-fats/

[2] 7 Things to Know About Omega-3 Fatty Acids. National Centre for Complementary and Integrative Health. Available from: https://www.nccih.nih.gov/health/tips/things-to-know-about-omega-fatty-acids

[3] Pan X-F, Marklund M, Wu JHY. Fish consumption for cardiovascular health: Benefits from long-chain omega-3 fatty acids versus potential harms due to mercury. British Medical Journal. 2019;**105**(18):1384-1385. DOI: 10.1136/heartjnl-2019-315187

[4] Zhang B, Xiong K, Cai J, Ma A. Fish consumption and coronary heart disease: A meta-analysis. Nutrients. 2020;**12**(8):2278

[5] Ganova-Iolovska M, Kalinov K, Geraedts M. Quality of care of patients with acute myocardial infarction in Bulgaria: A cross-sectional study. BMC Health Service Research. 2009;**9**(15):1472-6963. DOI: 10.1186/1472-6963-9-15

[6] Undeland I, Ellegård L, Sandberg A-S. Fish and cardiovascular health. Taylor & Francis Health Sciences. 2014;**48**(3):119-130. DOI: 10.1080/1102648041000042

[7] Harris WS. Fish oils and plasma lipid and lipoprotein metabolism in humans: A critical review. Journal of Lipid Research. 1989;**30**(6):785-807

[8] Kris-Etherton PM, Harris WS, Appel LJ. AHA nutrition committee American Heart Association. Omega-3 fatty acids and cardiovascular disease: New recommendations from the American Heart Association. Arteriosclerosis, Thrombosis, and Vascular Biology. 2003;**23**(2):151-152. DOI: 10.1161/01.atv.0000057393.97337.ae

[9] Marckmann P, Grönbäck M. Fish consumption and coronary heart disease mortality. A systematic review of prospective cohort studies. European Journal of Clinical Nutrition. 1999;**53**(8):585-590. DOI: 10.1038/sj.ejcn.1600832

[10] Marckman P. Fishing for heart protection. The American Journal of Clinical Nutrition. 2003;**78**(1):1-2. DOI: 10.1093/ajcn/78.1.1

[11] Popoff F, Balaciano G, Bardach A, Comandé D, Irazola V, Catalano HN, et al. Omega 3 fatty acid supplementation after myocardial infarction: A systematic review and meta-analysis. BMC Cardiovascular Disorders. 2019;**19**(1):136. DOI: 10.1186/s12872-019-1086-3

[12] Whelton SP, He J, Whelton PK, Muntner P. Meta-analysis of observational studies on fish intake and coronary heart disease. American Journal of Cardiology. 2004;**93**(9):1119-1123

[13] He K, Song Y, Daviglus M. Accumulated-evidence on fish consumption and coronary heart disease mortality. A meta-analysis of cohort studies. ACC Current. Journal Review. 2004;**109**(22):2705-2711. DOI: 10.1161/01.CIR.0000132503.19410.6B

[14] Zheng J, Huang T, Yu Y, Hu X, Yang B, Li D. Fish consumption and CHD

mortality: An updated meta-analysis of seventeen cohort studies. Public Health Nutrition. 2011;**15**(4):725-737. DOI: 10.1017/S1368980011002254

[15] Bechthold A, Boeing H, Schwedhelm C, Hoffmann G, Knüppel S, Iqbal K, et al. Food groups and risk of coronary heart disease, stroke and heart failure: A systematic review and dose-response meta-analysis of prospective studies. Critical Reviews in Food Science and Nutrition. 2017;**59**:1071-1090. DOI: 10.1080/10408398.2017.1392288

[16] Cassell DK. Why People at Risk of Heart Disease May Want to Avoid Fish Oil. 2021. Available from: https://www.healthline.com/health-news/why people at risk of heart disease may want to avoid fish oil

[17] Dariush Mozaffarian MD, Jason PH, Wu HY. Omega-3 fatty acids and cardiovascular disease: Effects on risk factors, molecular pathways, and clinical events. Journal of the American College of Cardiology. 2011;**58**(8):2047-2067. DOI: 10.1016/j.jacc.2011.06.063

[18] Søyland E, Funk J, Rajka G, et al. Effect of dietary supplementation with very-long-chain n-3 fatty acids in patients with psoriasis. New England Journal of Medicine. 1993;**25**:1812-1816

[19] Krauss RM, Eckel RH, Howard B, et al. AHA dietary guidelines: Revision. A statement for healthcare professionals from the Nutrition Committee of the American Heart Association. Circulation. 2000;**18**:2284-2299

[20] Mann NJ, O'Connell SL, Baldwin KM, Singh I, Meyer BJ. Effects of seal oil and tuna-fish oil on platelet parameters and plasma lipid levels in healthy subjects. Lipids. 2010;**45**(8):669-681.20652432

[21] Wardlaw GM. Perspectives in Nutrition. 4th ed. WBC/McGraw-Hill: Boston. MA; 1999

[22] Whitney EN, Rolfes SR. Understanding Nutrition. 8th ed. Belmont, CA: West/Wadsworth; 1999

[23] Bernasconi AA, Wiest MM, Lavie CJ, Milani RV, Laukkanen JA. Effect of omega-3 dosage on cardiovascular outcomes: An updated meta-analysis and meta-regression of interventional trials. Mayo Clinic Proceedings. 2020;**96**:304-313

[24] Lombardi M, Carbone S, Del Buono MG, et al. Omega-3 fatty acids supplementation and risk of atrial fibrillation: An updated meta-analysis of randomized controlled trials. European Heart Journal—Cardiovascular Pharmacotherapy. 2021;7(4):e69-e70. DOI: 10.1093/ehjcvp/pvab008

[25] Mozaffarian D, Rimm EB. Fish intake, contaminants, and human health: Evaluating the risks and the benefits. JAMA. 2006;**296**(15):1885-1899

[26] Mozaffarian D, Wu JH. (n-3) fatty acids and cardiovascular health: Are effects of EPA and DHA shared or complementary? The Journal of Nutrition. 2012;**142**(**3**):614S-6625S. DOI: 10.3945/jn.111.149633

[27] Pan A, Chen M, Chowdhury R. α-Linoleic acid and risk of cardiovascular disease: A systematic review and meta-analysis. The American Journal of Clinical Nutrition. 2012;**96**(6):1262-1273. DOI: 10.3945/ajcn.112.044040

[28] Djoussé L, Gaziano JM, Buring JE, Lee I-M. Dietary omega-3 fatty acids and fish consumption and risk of type 2 diabetes. The American Journal of Clinical Nutrition. 2011;**93**:143-150. DOI: 10.3945/ajcn.110.005603

[29] Kasim SE, Stern B, Khilnani S, McLin P, Baciorowski S, Jen KL. Effects of omega-3 fish oils on lipid metabolism,

glycemic control, and blood pressure in type II diabetic patients. The Journal of Clinical Endocrinology and Metabolism. 1988;**67**(1):1-5. DOI: 10.1210/jcem-67-1-1

[30] Friday KE, Childs MT, Tsunehara CH, Fujimoto WY, Bierman EL, Ensinck JW. Elevated plasma glucose and lowered triglyceride levels from omega-3 fatty acid supplementation in type II diabetes. Diabetes Care. 2017;**13**(1):80-81. DOI: 10.2337/diacare.13.1.80

[31] Woodman RJ, Mori TA, Burke V, Puddey IB, Watts GF, Beilin LJ. Effects of purified eicosapentaenoic and docosahexaenoic acids on glycemic control, blood pressure, and serum lipids in type 2 diabetic patients with treated hypertension. The American Journal of Clinical Nutrition. 2002;**76**(5):1007-1005

[32] Wang L, Folsom AR, Zheng ZJ, Pankow JS, Eckfeldt JH. Plasma fatty acid composition and incidence of diabetes in middle-aged adults: The Atherosclerosis Risk in Communities (ARIC) Study. The American Journal of Clinical Nutrition. 2003;**78**(1):91-98. DOI: 10.1093/ajcn/78.1.91

[33] Laaksonen DE, Lakka TA, Lakka HM, et al. Serum fatty acid composition predicts development of impaired fasting glycaemia and diabetes in middle-aged men. Diabetic Medicine. 2002;**19**(6):456-464. DOI: 10.1046/j.1464-5491.2002.00707.x

[34] Meyer KA, Kushi LH, Jacobs DR Jr, Folsom AR. Dietary fat and incidence of type 2 diabetes in older Iowa women. Diabetes Care. 2001;**24**(9):1528-1535. DOI: 10.2337/diacare.24.9

[35] Kaushik M, Mozaffarian D, Spiegelman D, Manson JE, Willett WC, Hu FB. Long-chain omega-3 fatty acids, fish intake, and the risk of type 2 diabetes mellitus. The American Journal of Clinical Nutrition. 2009;**90**(3):613-620. DOI: 10.3945/ajcn.2008.27424

[36] Hu FB, Manson JE, Stampfer MJ, et al. Diet, lifestyle, and the risk of type 2 diabetes mellitus in women. The New England Journal of Medicine. 2001;**345**(11):790-797. DOI: 10.1056/NEJMoa010492

[37] Hartweg J, Perera R, Montori V, Dinneen S, Neil HA, Farmer A. Omega-3 polyunsaturated fatty acids (PUFA) for type 2 diabetes mellitus. Cochrane database systemic. Review. 2008;**23**(1): CD003205. DOI: 10.1002/14651858.CD003205.pub2

[38] Bowman L, Mafham M, Wallendszus K, Stevens W, Buck G, Barton J, et al. Effects of n-3 fatty acid supplements in diabetes mellitus. New England Journal of Medicine. 2018;**379**(16):1540-1550. DOI: 10.1056/NEJMoa1804989

[39] Jacobo-Cejudo MG, Valdes-Ramos R, Guadarrama-Lopez AL, Pardo-Morales RV, Martinez-Carrillo BE, Harbige LS. Effect of n-3 polyunsaturated fatty acid supplementation on metabolic and inflammatory biomarkers in type 2 diabetes mellitus patients. Nutrients. 2017;**379**(16):1540-1550. DOI: 10.1056/NEJMoa1804989

[40] Brostow DP, Odegaard AO, Koh WP, Duval S, Gross MD, Yuan JM, et al. Omega-3 fatty acids and incident type 2 diabetes: The Singapore Chinese health study. American Journal of Clinical Nutrition. 2011;**94**(2):520-526. DOI: 10.3945/ajcn.110.009357

[41] Tajik B, Voutilainen A, Kauhanen J, Mazidi M, Lip GYH, Tuomainen TP, et al. Lipid profile, lipid ratios, apolipoproteins, and risk of cardiometabolic multimorbidity in men:

The Kuopio Ischaemic Heart Disease Risk Factor Study. European Publication. 2022;**57**(2):141-149

[42] Bernasconi AA, Wiest MM, Lavie CJ, Milani RV, Laukkanen JA. Effect of omega-3 dosage on cardiovascular outcomes: An updated meta-analysis and meta-regression of interventional trials. Mayo Clinic Proceedings. 2020;**96**(2):304-313. DOI: 10.1016/j.mayocp.2020.08.034

[43] Wu JH, Micha R, Imamura F, Pan A, Biggs ML, Ajaz O, et al. Omega-3 fatty acids and incident type 2 diabetes: A systematic review and meta-analysis. The British Journal of Nutrition. 2012;**2**:14-27. DOI: 10.1017/S0007114512001602

[44] Zoler ML. Oily fish linked to lower risk of diabetes in largest study to date. Web Med. 2021;**44**(3):672-680

[45] Feskens EJ, Virtanen SM, Rasanen L, et al. Dietary factors determining diabetes and impaired glucose tolerance. A 20-year follow-up of the Finnish and Dutch cohorts of the seven countries study. Diabetes Care. 1995;**18**:1104-1112

[46] Villegas R, Xiang YB, Elasy T, Li HL, Yang G, Cai H, et al. Fish, shellfish, and long-chain n-3 fatty acid consumption and risk of incident type 2 diabetes in middle-aged Chinese men and women. American Journal of Clinical Nutrition. 2011;**2**:543-551

[47] Brown TJ et al. Omega-3, omega-6, and total dietary polyunsaturated fat for prevention and treatment of type 2 diabetes mellitus: Systematic review and meta-analysis of randomised controlled trials. BMJ. 2019;**366**:l4697. DOI: 10.1136/bmj.l4697

[48] Hindricks G, Potpara T, Nikolaos Dagres N, et al. ESC guidelines for the diagnosis and management of atrial fibrillation developed in collaboration with the European Association of Cardio-Thoracic Surgery (EACTS). European Heart Journal. 2020;**42**(5):373-498. DOI: 10.1093/eurheartj/ehaa612

[49] Hjartaker A. Fish consumption and risk of breast, colorectal and prostate cancer: A critical evaluation of epidemiological studies. Scandinavian Journal of Nutrition. 2003;**47**(3):111-122. DOI: 10.1080/11026480310001959

[50] Bulliyya G. Fish intake and blood lipids in fish eating vs non-fish eating communities of coastal South India. Clinical Nutrition. 2000;**19**:165-170

[51] Marcovina SM, Kennedy H, Bon GB, Cazzolato G, Galli C, Caiglia E, et al. Fish intake, independent of Apo (a) size, accounts for lower plasma lipoprotein (a) levels in bantu fishermen in Tanzania. Atherosclerosis Thrombosis Vascular Biology. 1999;**19**:1250-1256

[52] Hu FB, Bronner L, Willet WC, Stampfler MJ, Rexrode KM, Albert CM. Fish and omega-3 fatty acid intake and risk of coronary heart disease in women. Journal of the American Medical Association. 2002;**287**:1815-1821

Section 3

Fatty Acid Metabolism and Cancer

Chapter 5

Fatty Acid Metabolism as a Tumor Marker

Gatot Nyarumenteng Adhipurnawan Winarno

Abstract

Cancer cells tend to make metabolism changes in the human body for their growth and survival. One of the most interesting changes is the alteration of fatty acid metabolism in order for the high rate of fatty acid synthesis required to increase the level of fatty acids needed for cancer cell proliferation. Thus, the reprogramming of fatty acid metabolism is needed for cancer cell survival. Fatty acid metabolic reprogramming is one of the hallmarks of the cancer condition since it can affect cellular functions. The reprogramming of fatty acid synthesis includes increased exogenous fatty acid uptake, de novo fatty acid synthesis, and oxidation of fatty acids. Identifying biochemical targets in fatty acid metabolism is useful for diagnosing and predicting the therapeutic efficiency in tumor treatment.

Keywords: fatty acids, cancer, metabolic reprogramming, tumor marker, fatty acid targeting

1. Introduction

Cancer cells need to control their environment to survive. Under a nutrient-limited microenvironment, they compete for oxygen and micronutrients with normal cells to survive and maintain their malignancy potential. Thus, oncogene-directed metabolic reprogramming is needed for cancer cell survival. Since lipids are essential molecules for cancer cells, understanding lipid metabolism reprogramming might be a useful hallmark for cancer. Those hallmarks include sustaining growth signaling, evading growth suppressors, resisting apoptosis programs, inducing angiogenesis, and activating invasion and metastasis.

In this chapter, we will focus on changes in fatty acid metabolism. Fatty acids are essential building blocks for some lipids with numerous essential roles in the human body. The body usually gains fatty acids from the diet, known as exogenous fatty acids. However, fatty acids can also be obtained endogenously through de novo fatty acid synthesis. In cancer conditions, a large number of fatty acids are needed. Changes in fatty acid metabolism are common to provide the needs of fatty acids for cancer cells.

A high rate of fatty acid metabolism must be followed by activating several enzymes involved in fatty acid metabolism. Overexpression of those enzymes might be useful as a marker for a cancer condition. Another benefit of understanding how a tumor cell influences the environment around it is also essential for developing new targeted therapies.

2. Fatty acids

Lipids are molecules that have numerous essential roles in the human body. In intracellular activity, lipids act as a ligand and second messenger. Lipids are composed of water-insoluble molecules such as fatty acids, triacylglycerides, phospholipids, or cholesterol. Fatty acids are molecules consisting of a carboxylic acid group and a hydrocarbon chain. Fatty acids are classified based on their hydrocarbon chain length: short-chain fatty acids (less than six carbon atoms), medium-chain fatty acids (6–12 carbon atoms), long-chain fatty acids (12–21 carbon atoms), and very-long-chain fatty acids (22 or more carbon atoms) [1–3].

Fatty acids are essential building blocks for lipids such as phospholipids, sphingolipids, and more complex lipids such as diacylglycerides (DAGs) and triacylglycerides (TAGs), which are important in regulating biochemical processes in a normal cell. In mammals, the main source of fatty acids is gained from the microenvironment (exogenous) through some transporter: fatty acid translocase (FAT) or cluster of differentiation 36 (CD36), fatty acid transport protein family (FATPs), and plasma membrane fatty-acid-binding proteins (FABPpm). In some conditions, fatty acids can be synthesized through de novo synthesis inside the body [3].

3. Synthesis and regulation of fatty acids

In order to build some essential lipids, fatty acids are obtained from direct uptake from the microenvironment. Exogenous fatty acids are taken up to the plasma membrane by some transporters: CD36, FATPs, and FABPpm. Those fatty acids are then stored as lipid droplets, which can convert to nicotinamide-adenine dinucleotide phosphate (NADPH) and acetyl CoA via β-oxidation, or known as fatty acid oxidation (FAO) [1, 3, 4].

NADPH is then used as energy fuel for metabolism activity, meanwhile acetyl-CoA is used as a material for the tricarboxylic acid (TCA) cycle. Acetyl-CoA is not only produced by β-oxidation, but is also generated from glucose and acetate metabolism. In the TCA cycle, acetyl-CoA is converted into citrate. When de novo fatty acid synthesis (FAS) is activated, those mitochondria-derived citrate together with glucose and amino acids then enter de novo FAS to produce fatty acids [3, 4].

De novo FAS is the formation of fatty acids from carbon atoms derived from carbohydrates such as glucose and amino acids, including glutamine. De novo synthesis occurs mainly in the liver and adipose tissue and is highly responsive to changes in dietary regimen. This pathway contributes to triacylglycerol's homeostasis, where most of the TAG is obtained from the diet. A high-carbohydrate diet activates a lipogenic response in liver tissue to increase the synthesis and secretion of very-low-density lipoprotein (VLDL) and increase hepatic de novo synthesis contributing to hypertriglyceridemia [1, 2].

De novo FAS requires some catalyst enzymes, such as ATP citrate lyase (ACLY), acetyl-CoA carboxylase (ACC), and fatty acid synthase (FASN). ACLY bridges glucose and FA metabolism by converting six-carbon citrate to oxaloacetate and two-carbon acetyl-CoA. The acetyl-CoA transforms into malonyl-CoA by ACC, which is regulated by citrate and glutamate. The next step produces 16-carbon palmitate from malonyl-CoA by FASN. Acetyl-CoA synthetase 2 (ACSS2) also plays a role in de novo FAS by converting acetate to acetyl-CoA before entering the TCA cycle [3, 5].

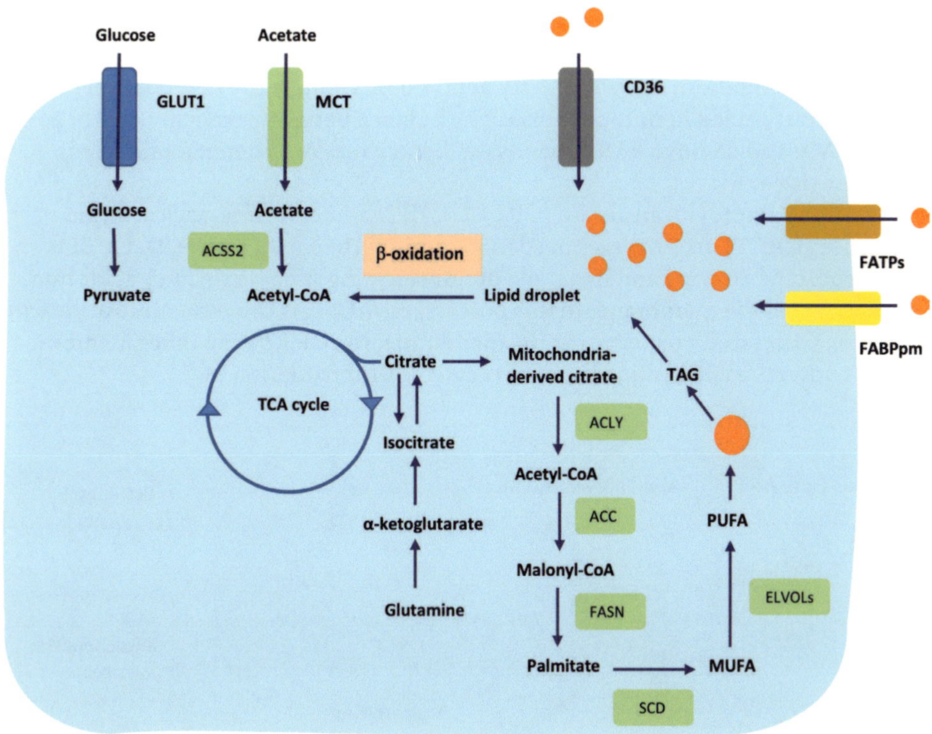

Figure 1.
Reprogramming of fatty acid metabolism in cancer cell [3]. GLUT1: glucose transporter 1, MCT: monocarboxylate transporter, CD36: cluster of differentiation 36, ACSS2: acetyl-CoA synthetase 2, TCA: tricarboxylic Acid, FATPs: fatty acid transport protein family, FABPpm: plasma membrane fatty-acid-binding proteins, ACLY: ATP citrate lyase, ACC: acetyl-CoA carboxylase, FASN: fatty acid synthase, SCD1: stearoyl-CoA desaturase-1, ELVOLs: elongation of very-long-chain fatty acid protein, TAG: triacylglycerides, PUFA: polyunsaturated fatty acids, MUFA: monounsaturated fatty acids.

One of TCA products is mitochondria-derived citrate, which converts to acetyl-CoA by ACLY in the cytosol. This acetyl-CoA is fuel for de novo FAS as it converts into malonyl-CoA by ACC. Finally, malonyl-CoA converts into palmitate by FASN (**Figure 1**). Subsequently palmitate is then esterified to form triglycerides or cholesterol esters. Palmitate may also form new fatty acids by elongation of very-long-chain fatty acid protein (ELOVLs) [1, 3, 4].

The regulation of de novo lipogenesis occurs through the activation of sterol regulatory element-binding proteins (SREBPs). There are three main transcription factors: SREBP1a and SREBP1c, and SREBP2. SREBPs are inactive 125-kDa precursors bound to the endoplasmic reticulum (ER). When the concentration of intracellular cholesterol is high, insulin-induced genes (INSIGs) will bind to SREBP-cleavage-activating proteins (SCAPs) and localize the SREBP precursors to the ER. But, when the cholesterol concentration is low, SCAPs will facilitate the translocation of ER-bound SREBPs to the Golgi and then release the active N terminus. This N-terminal fragment translocates to the nucleus and induces the transcription of genes containing sterol regulatory elements (SREs), such as FASN, ACLY, and ACC. Some growth factors and estrogen receptors also regulate FASN expressions, such as epidermal growth factor receptor (EGFR), human epidermal growth factor receptor 2 (HER2), estrogen receptor, and progesterone receptor [3].

4. Fatty acid oxidation (FAO)

Fatty acid oxidation, known as fatty acid β-oxidation, is the degradation of long-chain fatty acids in mitochondria. The balance between exogenous fatty acid uptake, FAO, and de novo FAS depends on dietary intake and metabolic rate in mitochondria [4, 6].

Carnitine palmitoyl transferase (CPT) 1, CPT2, and carnitine-acylcarnitine translocase (CACT) are enzymes needed to catalyze the FAO (**Figure 2**). CPT1 is an outer membrane enzyme and has a role in translocating long-chain fatty acids into acyl-CoA across the membrane. In this process, acyl-CoA is converted into acylcarnitine. After successfully penetrating the membrane, the CPT2 as an inner membrane enzyme converts acylcarnitine back into acyl-CoA for oxidation [4, 6].

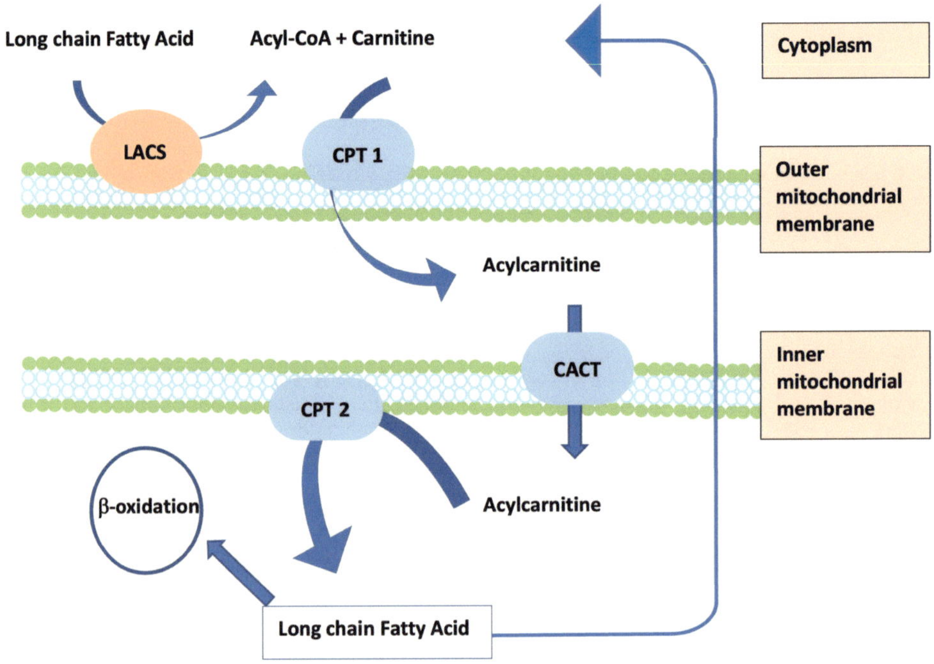

Figure 2.
Role of CPT1 and CPT2 in FAO [6]. LACS: long-chain acyl-coenzyme A (CoA) synthetases, CPT1: carnitine palmitoyl transferase 1, CPT2: carnitine palmitoyl transferase 2, CACT: carnitine-acylcarnitine translocase.

5. Reprogramming of fatty acid metabolism in cancer

Tumor cells need to control their environment to survive. Under a nutrient-limited microenvironment, they compete for oxygen and micronutrients with normal cells to survive and even maintain their malignancy potential. Changes in lipid metabolism are common to provide the needs of fatty acids for cancer cells. Cancer cells require fatty acids to build the membrane and molecule signaling. Thus, fatty acid metabolism reprogramming is one of the hallmarks of the cancer condition. This reprogramming includes alteration of fatty acid uptake, de novo fatty acid synthesis, and fatty acid oxidation (FAO) [3–5].

5.1 Increased exogenous fatty acid uptake

Overexpression of CD36 has been correlated with breast cancer survival and leads to elevation of exogenous fatty acid uptake. This elevated exogenous fatty acid uptake is implicated in enhancing tumor growth and metastasis and providing the cancer cell with chemoresistance. Elevated lipid droplets resulting from excess fatty acids are used by cancer cells for various vital roles, such as maintaining lipid homeostasis and preventing cancer cells from death under metabolic stress [3–5].

5.2 Increased de novo fatty acid synthesis

The body usually uses exogenous fatty acids to maintain homeostasis in normal conditions. De novo FAS is present when the diet is not adequate. But in the presence of cancer cells, increased de novo fatty acid synthesis usually occurs to meet the demand for high fatty acids. This is thought to be part of the general metabolic shift from a nonproliferative catabolic phenotype to a proliferative anabolic phenotype [2, 4].

Cancer cells upregulate some critical enzymes, such as ACC, ACLY, and FASN, to enhance the rate of de novo FAS (**Figure 3**). ACLY has a role in glucose metabolism and FA metabolism by converting six-carbon citrate to oxaloacetate and two-carbon acetyl-CoA. ACLY overexpression and activity have been represented in some cancers: in the lung, prostate, bladder, breast, and liver. This overexpression was correlated with a poorer prognosis in lung adenocarcinoma. Thus, inhibiting ACLY may induce activation of p53, which plays a role in suppressing cancer cell proliferation [3, 7].

FASN is a 270-kDa dimeric enzyme responsible for producing fatty acids from malonyl-CoA and acetyl-CoA. In breast cancer, HER2/neu increased FASN transcription and lipogenesis. In addition, a study found that HER2/neu directly phosphorylates FASN, and this phosphorylation increases FASN synthesis activity [2, 5, 8].

Overexpression of FASN has been known as a hallmark of phenotypic alteration for most human malignancies. This elevation of FASN leads to increase fatty

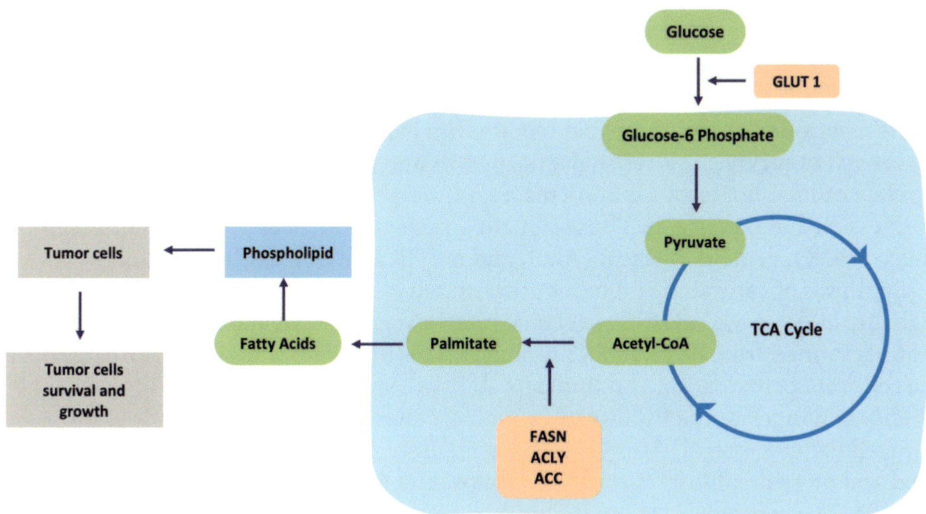

Figure 3.
De novo FA synthesis alteration. GLUT1: glucose transporter 1, FASN: fatty acid synthase, ACLY: ATP citrate lyase, ACC: acetyl-CoA carboxylase, TCA: tricarboxylic acid.

acid synthesis and correlates with poor prognosis for many cancer types. FASN also promotes tumor angiogenesis since it correlates with vascular endothelial growth factor (VEGF).

The level of fatty acids is also correlated with gynecological outcomes in epithelial ovarian cancer. A study reports that a higher level of fatty acids is correlated with suboptimal debulking in epithelial ovarian cancer patients. Increased FASN expression and activity were also found in early oncogenesis and correlated with cancer development. A study in ovarian cancer showed that overexpression of FASN is more correlated with tumor proliferation than tumor metastasis. Conversely, FASN inhibition leads to accumulating malonyl-CoA, which induces apoptosis [3, 5, 9, 10].

Cancer cells tend to upregulate ACC under metabolic stress or lipid depletion. These ACC then convert acetyl-CoA to malonyl-CoA to meet the high demand for fatty acids. On the other hand, ACC inhibition decreases palmitic acid and induces an apoptosis program. In a breast cancer cell, apoptosis triggered by depletion of ACC only occurs in the nonmalignant cell [3, 10].

5.3 Increased fatty acid oxidation

Another reprogramming of fatty acid metabolism is increased fatty acid oxidation. The products of fatty acid oxidation such as NADH, NADPH, FADH2, and ATP are used for cancer cells as energy-carrying molecules, which are important for their survival. High levels of ATP from FAO are correlated with the proliferation of triple-negative breast cancer. Acetyl-CoA from the FAO also can be used by cancer cells to gain fatty acids through de novo FAS. An increase in CPT1 is common in cancer cells. CPT1 also provides neovascularization for the tumor microenvironment [4, 6].

CPT1C is one of the subtypes of CPTC, which is upregulated in some cancers and protects the cancer cells from metabolic stress such as hypoxia or glucose depletion. Elevation of CPT1C in cancer cells provides FAO and adaptation to metabolic stress and resistance to mTOR complex 1 (mTORC1) inhibitor. Thus, increased FAO is essential not only for tumor growth but also for cancer chemoresistance [4, 6].

6. The role of fatty acids for tumorigenesis

As mentioned above, fatty acid metabolism has some essential roles in helping the cancer cell to survive. Interestingly, fatty acids are involved in signaling pathways to regulate cellular homeostasis and create an appropriate microenvironment for tumor progression and metastasis. Those signaling pathways include phosphoinositide 3 kinase (PI3K), protein kinase B (Akt), and mTOR signaling pathways [3, 11].

Synthesis of saturated and monounsaturated fatty acids is increased due to the high rate of de novo FAS in cancer cells. Interestingly, a rigid and stable cell membrane is formed from de novo FAS products. This high cholesterol structure protects the cell from peroxidation and limits its ability to spread (**Figure 4A**). To reduce the rigidity and increase the fluidity of cell membranes, the cancer cell needs to release intracellular cholesterol. After decreasing cholesterol levels, the cell membrane is less rigid and more flexible in changing its shape, and epithelial-to-mesenchymal transition (EMT) can occur [3, 11].

EMT is an important mode for cancer cells to spread. As an enzyme involved in de novo FAS, FASN correlates with EMT. The level of FASN is increased in EMT cells. FASN can regulate the Wnt or (transforming growth factor β) TGF-β signaling

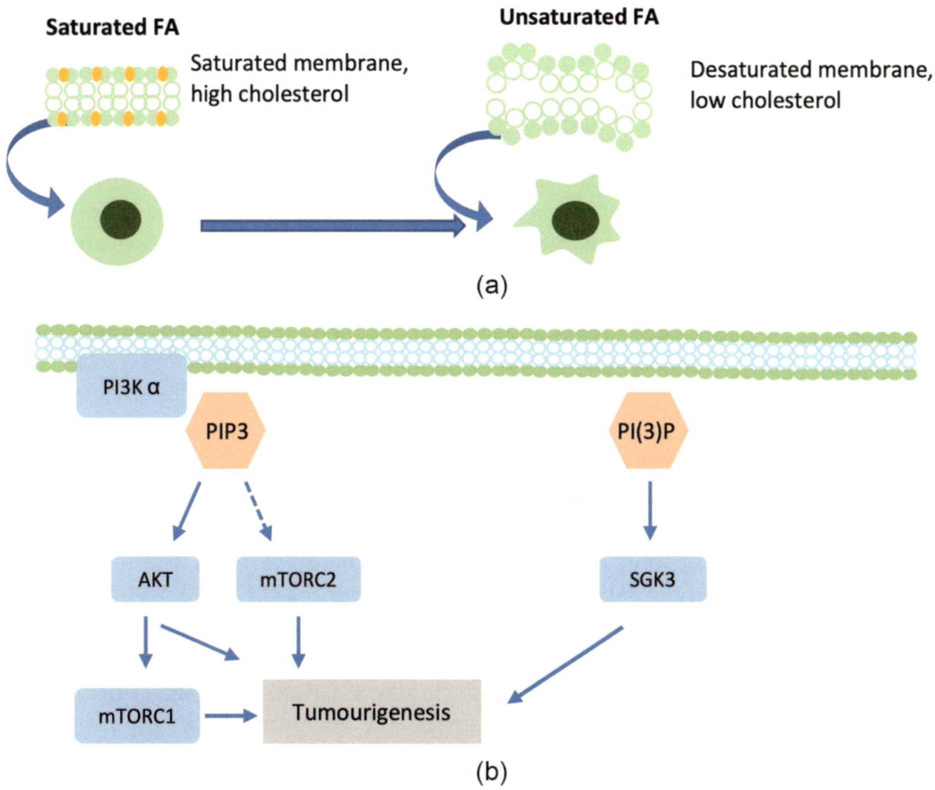

Figure 4.
Role of fatty acids in signaling pathway [3]. PI3Kα: phosphoinositide 3-kinase α, PIP3: phosphatidylinositol (3,4,5)-trisphosphate, PI(3)P: phosphatidylinositol 3-phosphate, AKT: protein kinase B, mTORC: mammalian target of rapamycin complex, SGK3: serum and glucocorticoid-induced protein kinase-3.

pathways to alter cell membrane rigidity and increase the mobility of cancer cells. This opportunity gives the cancer cell ability to spread. But conversely, high cholesterol levels in cell membrane lead to cholesterol-rich lipid rafts formation and induce oncogenic signaling via PI3K-AKT (**Figure 4B**). With this contradictive theory, inhibiting cholesterol synthesis still can be considered as a therapy for early-stage cancer [3, 11].

Another signaling molecule, which contributes in tumorigenesis, is phosphatidylinositol (PtdIns). PtdIns is composed of two fatty acid chains bound to an inositol ring and glycerol. Some phosphoinositide species, such as phosphatidylinositol (3,4)-bisphosphate or PI(3,4), phosphatidylinositol (4,5)-bisphosphate or PI(4,5), phosphatidylinositol 3-phosphate or PI(3)P, and phosphatidylinositol (3,4,5)-trisphosphate or PIP3, are generated from hydroxyl group of inositol ring. PIP3 activates oncogenic AKT and induces activation of mammalian target of rapamycin complex 2 (mTORC2), leading to tumorigenesis. On the other hand, PI(3)P induces tumorigenesis via serum- and glucocorticoid-induced protein kinase-3 (SGK3) activation.

7. Targeting fatty acid metabolism for cancer treatment

Considering reprogramming fatty acid metabolism is an important hallmark for cancer development; targeting key enzymes involved in fatty acid metabolism is a promising strategy. Not only de novo FAS and FAO, but lipid uptake is also a potential target for cancer treatment.

7.1 Targeting lipid uptake

CD36 inhibition might be useful for tumors with lipoprotein lipase (LPL) and CD36 overexpression. Anti-CD36 use implicates complete inhibition of metastasis in immunodeficient mice without any adverse effect. In prostate cancer, CD36 monoclonal antibody use can reduce exogenous fatty acid uptake and several oncogenic signaling lipids. Further study is still needed to reveal the side effects of long-term inhibition of CD36. Decreased lipid uptake into the cell using orlistat may also be used for targeted therapy. Orlistat is used to inhibit LPL in order to decrease exogenous fatty acid uptake [4].

7.2 Targeting enzymes involved in de novo FAS

Since the role of FASN is important in supporting both anabolic metabolism and oncogenic signaling, targeting FASN for cancer therapy might be promising. Inhibition of FASN leads to toxic accumulation of malonyl-CoA in inducing cell death. In cells derived from a lymph node metastasis of prostate carcinoma (LNCaP), RNA interference (RNAi) activity against FASN implicates cancer cell growth inhibition. First-line chemical inhibitors of FASN such as C75, orlistat, and cerulenin are shown to reduce and inhibit tumor growth by inducing cell-cycle arrest. Orlistat provides antitumor properties for some cancers such as breast cancer and melanoma. FASN inhibition was also sensitive for ovarian cancer cells by C75 and G28UCM. Using this first generation of FASN inhibitors had some adverse effects such as increased energy expenditure, loss of adipose tissue, and decreased body weight [3–5].

The next generation of FASN inhibitors, such as TVB-3166 and TVB-2640, has shown antitumor potential, higher specificity for FASN, and limited systemic toxicity in a preclinical study. Antitumor activity has been shown in breast cancer and colorectal cancer by TVB-3166 and TVB-3664 use in preclinical study. In a clinical trial, TVB- 3166 and TVB-2640 showed limiting systemic toxicity in early phase. The significant difference between the first and next generations is that the newer generation does not implicate indirect CPT1 in peripheral tissue. Omeprazole may also be used for FASN inhibitors and has entered clinical trials in triple-negative breast cancer patients [3, 4].

Some ACLY inhibitors have shown high efficacy in lowering LDL cholesterol. ETC- 1002 has entered clinical trial in phase 2/3, and hydroxy citrate has already entered randomized control trials. Thus, ACLY inhibitors can be considered as a therapeutic strategy for cancer therapy and have been demonstrated to reduce tumor proliferation in some cancers, such as breast cancer, lung cancer, and prostate cancer [4].

SCD1 inhibitors such as CVT 11127, MF-483 also have potential as an anticancer therapeutic target. The combination of SCD1 inhibitor with other drugs shows potency against cancer cell chemoresistance and enhances therapeutic efficacy. Commonly used drugs in combination with SCD1 are gefitinib, temozolomide, and temsirolimus [4].

AKT/mTOR/SREBP-1 pathway is also interesting to point out as targeted therapy. The activity of fatty acid synthase may be reduced by reducing its transcription levels. Inhibiting SREBP-1 in cancer cells could decrease fatty acid synthase gene expression and prevent cancer cell growth. Among SREBP-1 inhibitors, betulin has increased the sensitivity of hepatocellular carcinoma cells to their first-line anti-tumor agent, sorafenib. On the other hand, a combination of the mTOR inhibitor everolimus and metformin can inhibit the proliferation of breast cancer cells [4].

7.3 Targeting fatty acid catabolism

Limiting excess fatty acids by increasing fatty acid degradation by mitochondria β-oxidation can be beneficial in reducing the proliferation of cancer cells. Otherwise, the high rate of fatty acid oxidation could prevent cancer cells from metabolic stress by providing abundant ATP as an energy fuel. Thus, decreasing fatty acid oxidation might be a promising strategy against cancer cells [4, 5].

CPT1 is the first and rate-limiting step of fatty acid oxidation. The use of CPT1 inhibitors such as etomoxir may inhibit tumor cell proliferation. Perhexiline, aβ-oxidation inhibitor has also shown to be beneficial by inhibiting fatty acid

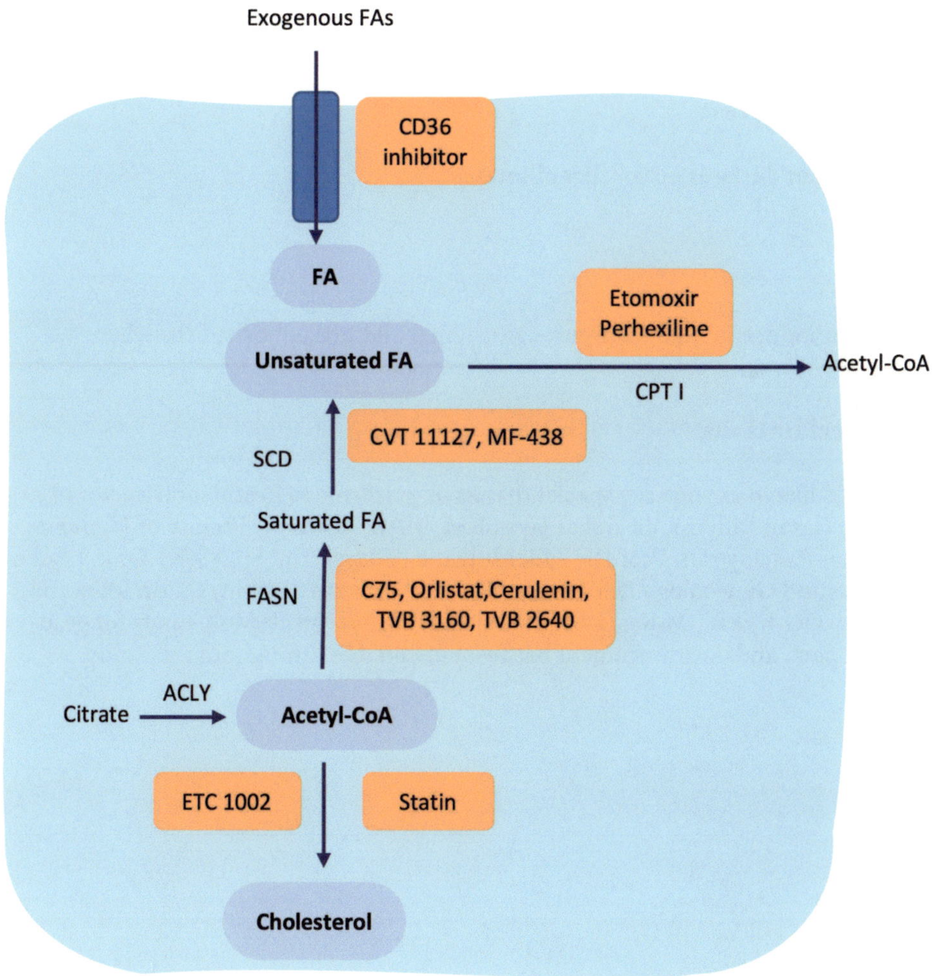

Figure 5.
Therapuetic targets in fatty acid metabolism [4]. CD36: cluster of differentiation 36, SCD: stearoyl-CoA Desaturase, FASN: fatty acid synthase, ACLY: ATP citrate lyase, CPT1: carnitine palmitoyl transferase-1.

utilization and tumor cell proliferation. In addition to perhexiline, sensitivity in gastrointestinal cancer for its first-line regimen therapy has increased (**Figure** 5) [4].

8. Conclusions

Fatty acids are essential building blocks for lipids, which have numerous essential roles in the human body. Fatty acids are needed for cell survival, not only in healthy cells but also in cancer cells. Thus, fatty acid reprogramming is a useful hallmark for cancer conditions. Understanding several enzymes involved in fatty acid metabolism, such as FASN, ACC, and ACLY, is also helpful in developing therapeutic targets for cancer.

Acknowledgements

This study was supported by a grant from the Department of Obstetrics and Gynecology, RSUP Dr. Hasan Sadikin and Universitas Padjajaran Bandung.

Conflict of interest

The author declares no conflict of interest.

Pictures

The author declares all the figures as original and not published elsewhere.

Other declarations

I would like to express my special thanks of gratitude to President Director of RSUP Dr. Hasan Sadikin, dr. Azhar Jaya SKM, MARS, Dean of Faculty of Medicine Universitas Padjdjaran, Prof Dr. Yudi Mulyana Hidayat, dr., SpOG(K)-Onk, head of Obstetrics and Gynecology Department RSUP Dr. Hasan Sadikin, Dr. dr. Ruswana Anwar, Sp.OG-KFER., M.Kes, and my assistant Dr Ayu Insafi Mulyantari for guidance, support, and encouragement has been invaluable throughout this study.

Author details

Gatot Nyarumenteng Adhipurnawan Winarno
Faculty of Medicine, Department of Obstetrics and Gynecology, Universitas Padjadjaran, Bandung, Indonesia

*Address all correspondence to: gatotnaw@yahoo.com

References

[1] Matsushita Y, Nakagawa H, Koike K. Lipid metabolism in oncology: Why it matters, how to research, and how to treat. Cancers. 2021;**13**(3):474

[2] Baumann J, Sevinsky C, Conklin DS. Lipid biology of breast cancer. Biochimica et Biophysica Acta (BBA) - Molecular and Cell Biology of Lipids. 2013;**1831**(10):1509-1517

[3] Koundouros N, Poulogiannis G. Reprogramming of fatty acid metabolism in cancer. British Journal of Cancer. 2020;**122**(1):4-22

[4] Wang W, Bai L, Li W, Cui J. The lipid metabolic landscape of cancers and new therapeutic perspectives. Frontiers in Oncology. 2020;**10**:605154

[5] Currie E, Schulze A, Zechner R, Walther TC, Farese RV. Cellular fatty acid metabolism and cancer. Cell Metabolism. 2013;**18**(2):153-161

[6] Qu Q, Zeng F, Liu X, Wang QJ, Deng F. Fatty acid oxidation and carnitine palmitoyltransferase I: Emerging therapeutic targets in cancer. Cell Death & Disease. 2016;**7**(5):e2226-e2226

[7] Zaidi N, Swinnen JV, Smans K. ATP-citrate lyase: A key player in cancer metabolism. Cancer Research. 2012;**72**(15):3709-3714

[8] Liu H, Liu JY, Wu X, Zhang JT. Biochemistry, molecular biology, and pharmacology of fatty acid synthase, an emerging therapeutic target and diagnosis/prognosis marker. International Journal of Biochemistry and Molecular Biology. 2010;**1**(1):69-89

[9] Winarno G, Hidayat Y, Soetopo S, Krisnadi S, Tobing M, Rauf S. Higher level of fatty acid synthase enzyme predicts lower rate of completing debulking surgery in epithelial ovarian cancer. Asian Pacific Journal of Cancer Prevention. 2020;**21**(10):2859-2863

[10] Mashima T, Seimiya H, Tsuruo T. De novo fatty-acid synthesis and related pathways as molecular targets for cancer therapy. British Journal of Cancer. 2009;**100**(9):1369-1372

[11] Yu XH, Ren XH, Liang XH, Tang YL. Roles of fatty acid metabolism in tumourigenesis: Beyond providing nutrition (Review). Molecular Medicine Report. Dec 2018;**18**(6):5307-5316. DOI: 10.3892/mmr.2018.9577. Epub 2018 Oct 23. PMID: 30365095

Chapter 6

An Important Component of Tumor Progression: Fatty Acids

Jin Wang, Qifei Wang and Guangzhen Wu

Abstract

Fatty acids (FAs) are complex and essential biomolecules in the human body and are critical to the formation of cell membranes, energy metabolism, and signaling. FAs are the major components of several lipids including phospholipids, sphingolipids, and triglycerides, and consist of carboxylic acid groups and hydrocarbon chains of different carbon lengths and degrees of desaturation. They can synthesize more complex lipids, including acylglycerides (DAG) and triacylglycerides (TAG). Saturated fatty acids (SFA), polyunsaturated fatty acids (PUFA), and monounsaturated fatty acids (MUFA) can be classified according to whether the hydrocarbon chain is saturated or not. Normal cells are commonly supplied with energy by the tricarboxylic acid cycle. On the contrary, to obtain energy, tumor cells usually use aerobic glycolysis (Warburg effect) and produce large amounts of FAs to maintain membrane structure to support cell proliferation. In addition, cancer migration, immune escape, development of drug resistance, and fatty acids are very closely related. In conclusion, a deeper understanding of the molecular mechanisms of fatty acid metabolism could provide a more plausible explanation for the progression of cancer cells and provide new potential targets for therapy.

Keywords: fat acids, tumor progression, cancer, lipid mediators, fatty acid metabolism

1. Introduction

FAs are important for human health and nutrition and are characterized by a wide variety and functional complexity. We have summarized some of the main functions of FAs as follows.

1. Provision of energy: Fatty acids are a major component of triglyceride synthesis. On the one hand, fatty acids can be esterified with glycerol for efficient energy storage [1], and on the other hand, it can also produce ATP through fatty acid oxidation (FAO, also known as β-oxidation) to supply energy to the body. FAO is a multistep catabolic process. Long-chain fatty acids are first converted to acetyl-CoA, which is then fully oxidized by the tricarboxylic acid (TCA) cycle and the electron transport chain (ETC), producing ATP. Several studies have shown that abnormal activity of FAO is associated with multiple aspects of tumorigenesis. Many cancer cells are FAO-dependent for proliferation, survival,

drug resistance, or metastasis. So, the study of key enzymes and regulatory mechanisms in the FAO process is helpful for cancer treatment [2].

2. Constituting biological membranes: Acetyl-CoA is recycled back to fatty acyl CoA, then converted to fatty acids via fatty acid synthase (FAS) [3]. Fatty acids are prolonged and desaturated by stearoyl coenzyme A desaturase (SCD1) to produce some of the unsaturated fatty acids, triglycerides, and phospholipids required for the synthesis of biological membranes [4]. Both these and cholesterol are essential components of biological membranes. The biofilm is a functionally complex structure composed of thousands of different lipids, and all communication inside and outside the cell is controlled by it [5].

3. Synthesis of signaling molecules as lipid mediators. For example, arachidonic acid (AA) is an omega-6-derived PUFA that can synthesize eicosanoids, including prostaglandins and thromboxanes, and can also serve as a substrate for leukotriene synthesis via the lipoxygenase pathway. Prostaglandins play a role in inhibiting inflammation and promoting a tumorigenic environment [6].

2. De novo synthesis of fatty acids

Most adult mammalian cells obtain lipids from the blood, which are partly derived from the diet and partly from hepatic synthesis. Still, cancer reactivates de novo lipogenesis, making them more independent of externally supplied lipids themselves. Numerous research results have verified that the growth and survival of cancer cells depend on the de novo synthesis of FA [1, 7]. De novo fatty acid biosynthesis in the adult organism occurs primarily in the liver, adipose tissue, and lactating breast [8]. In cells, glucose, glutamine, or acetate produce acetyl-CoA, which synthesizes FAs [9]. Acetyl-CoA is activated by acetyl-CoA carboxylase (ACC) to form malonyl CoA; it is then condensed with the participation of FAS to form the 16 carbon saturated fatty acid palmitate. Next, palmitate undergoes elongation and desaturation to create a cellular pool of nonessential FAs, which include the 18 carbon monounsaturated fatty acid oleate (C18:1). Multiple oncogenic signaling pathways are focused on FA synthesis. For example, the PI3K/Akt signaling pathway, on the one hand, promotes the expression of enzymes required for FA synthesis. On the other hand, it increases the phosphorylation and activation of ATP citrate lyase (ACLY). ACLY is a key enzyme linking glucose metabolism and lipid synthesis and can catalyze cytoplasmic citrate to acetyl-CoA [10, 11]. In contrast, under the STK11/LKB1 tumor suppressor pathway control, AMP-activated protein kinase (AMPK) blocks FA synthesis by phosphorylating ACC [12]. Because increased de novo FA synthesis in cancer cells alters cellular lipid composition, it can be used for diagnosis [13]. Also, FA uptake is essential for cancer, and reducing fatty acid uptake of prostate cancer cells by silencing CD36 has been reported for use in preclinical models for the treatment of prostate cancer [14].

3. Function of fatty acids in cancer cells

The tumor tissue must also acquire some lipids from the extracellular environment. The uptake and function of FAs require the involvement of fatty acid-binding protein (FABP). FABPs are a family of proteins involved in FA uptake and transport,

FABPS	Characteristics	Refs
FABP1	FABP1 is also known as liver-FABP, is widely distributed in the cytoplasm of hepatocytes and is less distributed in the nucleus and the outer mitochondrial membrane.	[15]
FABP2	FABP2 is also known as intestinal-FABP, is expressed in the epithelium of the small intestine.	[16]
FABP3	FABP3, which preferentially binds to n-6 PUFAs such as AA, is expressed in the brain, heart, skeletal muscle, lactating mammary glands, and placenta.	[17]
FABP4	FABP4 may be secreted from adipocyte cells and transferred to tumor cells. In breast and prostate cancer, FABP4 promotes cell growth and metastasis.	[18, 19]
FABP5	FABP5 is upregulated in a variety of cancers, including triple-negative breast cancer, bladder cancer, pancreatic cancer, and oral squamous cell carcinoma.	[20]
FABP6	FABP6 is mainly expressed in ileal epithelial cells, which plays an extremely important role in the intracellular transporter of bile acids and the metabolism of cholesterol.	[21, 22]
FABP7	FABP7 is well known for its significant expression in brain tissue and its function has been extensively studied in glioblastoma cell lines.	[23]
FABP8	FABP8 is a major cytosolic protein constituent of myelin in the peripheral nervous system.	[24]
FABP9	FABP9 is a member of the FABPs family and is mainly expressed in testis tissue.	[25]

Table 1.
Characteristics of FABP family proteins.

and nine FABP factors have been identified in mammalian cells and are numbered in the order of their discovery (FABP1, FABP2, FABP3, FABP4, FABP5, FABP6, FABP7, FABP8 and FABP9) (**Table 1**).

3.1 Promotes tumor growth

Due to the active proliferative characteristics of tumor cells, FA is required to provide the lipids and energy necessary to form cellular structures, so cell growth must be closely related to the induction of lipid synthesis. Moreover, a common feature of cancer is the conversion to a glucose-dependent form of metabolism. Once the glucose metabolism of tumor cells exceeds the bioenergetic requirements, the excess metabolites are combined into lipids. For example, inhibition of ACLY with small molecules can impair the development of immortalized hematopoietic stem cells stimulated by the corresponding growth factors [26]. There are other studies showing that ACLY overexpression promotes tumor cell growth, while inhibition of ACLY inhibits tumor cell growth [27–29].

3.2 Changing the composition of membrane lipids

It is also possible to change the composition of lipids in biological membranes to affect the function of organelles in membrane-containing cells, thereby altering FA synthesis and modification. Cardiolipids (CLs) are structurally specialized phospholipids located mainly in the inner mitochondrial membrane, where they control mitochondrial respiration and can signal the induction of apoptotic processes [30]. Regulation of FA biosynthesis and uptake in cancer cells can directly influence cellular bioenergetics by modulating the electron transport chain (ETC) activity. It was shown that the CL profile of mouse brain tissue, both in mitochondria separated from tumor tissue and normal tissue, was significantly different and correlated with impaired ETC enzyme activity [31]. Therefore, we can assume that interfering with CL synthesis, and thus mitochondrial function, has the potential to have a therapeutic effect on cancer.

3.3 Protein acylation

The abundance and degree of saturation of cellular FA also dictate that signaling protein activity requires acylation to function. For example, WNT (wingless-related MMTV integration site) proteins are a family of proteins that regulate development, and its dysregulation is closely associated with many disease-related processes and cancers [32]. Disruption of epithelial cell polarity due to aberrant activation of the WNT-β-catenin pathway leads to loss of cell–cell adhesion [33]. SRC and RAS oncoproteins are also cancer-associated acylated proteins modified mainly through thioesterification of saturated palmitoyl residues [34].

3.4 Lipid mediators

Lipid mediators are bioactive molecules derived from PUFAs, such as AA, docosahexaenoic acid (DHA), and eicosapentaenoic acid (EPA), through enzymatic or non-enzymatic oxidative metabolism [35]. For example, these compounds include sphingosine-1-phosphate (S1P) [36] and lysophosphatidic acid (LPA) [37], which are biologically active lipids. Diacylglycerol (DAG) [38], inositol-1,4,5-trisphosphate (IP3) and phosphatidylinositol-3,4,5-trisphosphate (PIP3) [39], can act as lipid second messengers. S1P is a member of the sphingolipid family and is involved in the regulation of various physiological responses, including cell growth, transformation, migration, and cell death [40]. LPA is signaled through autocrine and paracrine mechanisms and has multiple effects on cell proliferation, differentiation and intracellular messaging, stimulating cell migration, inflammation, and angiogenesis [37]. Targeting lipid mediators in the tumor microenvironment or within the tumor may be a new therapeutic approach for treating cancer patients.

4. Role of fatty acids in cancer progression

Although cancers are a series of distinct diseases, there are common features among cancers such as proliferation, apoptosis resistance, metabolic adaptation, migration, and invasion, and these features contribute to cancer progression. Fatty acids not only have an essential factor in the development of cancer but also play a crucial role in tumor development and metastasis, and even become a major cause of cancer death.

4.1 Cell migration

Usually, epithelial-mesenchymal transition (EMT) leads to increased motility of cancer cells [41]. EMT consists of multiple dynamic transition states between epithelial and mesenchymal phenotypes that can regulate tumor progression and metastasis [42]. Induction of EMT can promote changes in membrane fluidity required for cell migration by remodeling the composition of cellular lipids. After tail vein injection in mice, breast cancer cells can be treated with compounds that disrupt the gene expression profile associated with EMT to reduce membrane fluidity and block cancer cell migration and lung metastasis formation [43]. Cancer cells may change from a state of proliferation to a state of migration in such a way that FA uptake or selective release of specific FA species from membrane lipids creates a signaling molecule that promotes cell migration and invasion.

4.2 Angiogenesis

Normally, angiogenesis occurs only during embryonic development, the female reproductive cycle, and wound repair. While aberrant angiogenesis is a key mediator and major process in cancer development [44]. The induction of angiogenesis determines the proliferation of metastasis and the growth of primary and metastatic tumors [45]. Lipid signaling, including prostaglandin E2 (PGE2), lysophosphatidic acid (LPA), and sphingosine-1-phosphate (S1P), can stimulate blood vessel growth while recruiting immune cells (especially macrophages), promoting tumor angiogenesis [46, 47]. As the possibilities of anti-angiogenic therapy are continually explored, it will become possible to target cancer with angiogenesis.

4.3 Immune escape

Immune editing refers to how cancer cells escape immune surveillance through evolution. Cancer cells can suppress the cytotoxic function of T cells by expressing checkpoint proteins and changing macrophages into a pre-tumorigenic phenotype by reprogramming them. The interactions between cancer and stromal cells involved in immune editing are complex and these interactions may be regulated by lipid-derived factors [48], such as PEG2.

5. Utilization of fatty acid metabolism

In countries around the world, cancer is the leading cause of death. The burden of cancer incidence and mortality is rapidly increasing worldwide, so the treatment of cancer has been a focus of clinical development [49, 50]. There is growing evidence that lipid metabolic reprogramming is present in a variety of cancers [7, 51]. Because of the close relationship between FA metabolism and cancer pathogenesis, it is clinically significant to develop therapeutic approaches that target FA metabolic reprogramming. Not surprisingly, FAS has received the most attention in tumor-targeted therapies associated with dysregulated lipid metabolism, due to its multifaceted functions in supporting anabolic and oncogenic signaling. It has been shown that when comparing normal tissues with malignant ovarian cancer models, the expression of FAS mainly reflects the proliferation and growth status of cells rather than the degree of malignancy [52]. Therefore, FAS inhibitors are widely concerned with cancer therapy, such as the natural product cerulenin and the new compound C75, which selectively kill cancer cells by inducing apoptosis [53]. Despite these encouraging results, the issue of potential side effects of targeted FAS remains serious. Another validated target is ACLY. Previous studies have observed increased ACLY expression and activity in glioblastoma, colorectal, breast, and hepatocellular carcinomas [26]. There is substantial preclinical evidence to support the overall role of ACLY in tumorigenesis, not only for its genetic targeting but also for its pharmacological targeting that significantly reduces the growth of xenografts in lung and prostate tumors and this antitumor effect is more pronounced in highly glycolytic cells [26].

6. Conclusion

Past studies have shown that lipid metabolism, especially FA synthesis, plays an important role in the normal human body. Alterations in FAs not only affect cancer cell migration but also induce angiogenesis, metabolic symbiosis, immune evasion, and drug resistance. FA is similarly important in the progression of cancer cells and FA synthesis has received much attention as a potential target for cancer therapy. However, it seems that clinical practice is still scarce. The reason for this may be that it is difficult for us to inhibit lipid metabolism in cancer cells in a highly selective manner without affecting other tissues throughout the body. Moreover, cellular lipids are diverse, complex in function, and also closely related to other metabolisms, so targeting this aspect of cancer cell metabolism remains challenging. Therefore, the link between FA and tumors requires further research. The development of drug resistance has been one of the main reasons for high cancer-related mortality, and alterations in lipid metabolism can also affect drug resistance. This also provides new strategies to explore cancer treatment.

Author details

Jin Wang, Qifei Wang* and Guangzhen Wu*
Department of Urology, The First Affiliated Hospital of Dalian Medical University, Dalian, China

*Address all correspondence to: wuguang0613@hotmail.com
and wangqifei6008@hotmail.com

References

[1] Santos CR, Schulze A. Lipid metabolism in cancer. The FEBS Journal. 2012;**279**(15):2610-2623. DOI: 10.1111/j.1742-4658.2012.08644.x

[2] Caro P, Kishan AU, Norberg E, et al. Metabolic signatures uncover distinct targets in molecular subsets of diffuse large B cell lymphoma. Cancer Cell. 2012;**22**(4):547-560. DOI: 10.1016/j.ccr.2012.08.014

[3] Wakil SJ. Fatty acid synthase, a proficient multifunctional enzyme. Biochemistry. 1989;**28**(11):4523-4530. DOI: 10.1021/bi00437a001

[4] Enoch HG, Catalá A, Strittmatter P. Mechanism of rat liver microsomal stearyl-CoA desaturase. Studies of the substrate specificity, enzyme-substrate interactions, and the function of lipid. The Journal of Biological Chemistry. 1976;**251**(16):5095-5103

[5] Enkavi G, Javanainen M, Kulig W, Róg T, Vattulainen I. Multiscale simulations of biological membranes: The challenge to understand biological phenomena in a living substance. Chemical Reviews. 2019;**119**(9):5607-5774. DOI: 10.1021/acs.chemrev.8b00538

[6] Wang D, Dubois RN. Eicosanoids and cancer. Nature Reviews. Cancer. 2010;**10**(3):181-193. DOI: 10.1038/nrc2809

[7] Currie E, Schulze A, Zechner R, Walther TC, Farese RV Jr. Cellular fatty acid metabolism and cancer. Cell Metabolism. 2013;**18**(2):153-161. DOI: 10.1016/j.cmet.2013.05.017

[8] Menendez JA, Lupu R. Fatty acid synthase and the lipogenic phenotype in cancer pathogenesis. Nature Reviews. Cancer. 2007;**7**(10):763-777. DOI: 10.1038/nrc2222

[9] Pietrocola F, Galluzzi L, Bravo-San Pedro JM, Madeo F, Kroemer G. Acetyl coenzyme a: A central metabolite and second messenger. Cell Metabolism. 2015;**21**(6):805-821. DOI: 10.1016/j.cmet.2015.05.014

[10] Bauer DE, Hatzivassiliou G, Zhao F, Andreadis C, Thompson CB. ATP citrate lyase is an important component of cell growth and transformation. Oncogene. 2005;**24**(41):6314-6322. DOI: 10.1038/sj.onc.1208773

[11] Berwick DC, Hers I, Heesom KJ, Moule SK, Tavare JM. The identification of ATP-citrate lyase as a protein kinase B (Akt) substrate in primary adipocytes. The Journal of Biological Chemistry. 2002;**277**(37):33895-33900. DOI: 10.1074/jbc.M204681200

[12] Shackelford DB, Shaw RJ. The LKB1-AMPK pathway: Metabolism and growth control in tumour suppression. Nature Reviews. Cancer. 2009;**9**(8):563-575. DOI: 10.1038/nrc2676

[13] Hilvo M, Denkert C, Lehtinen L, et al. Novel theranostic opportunities offered by characterization of altered membrane lipid metabolism in breast cancer progression. Cancer Research. 2011;**71**(9):3236-3245. DOI: 10.1158/0008-5472.CAN-10-3894

[14] Watt MJ, Clark AK, Selth LA, et al. Suppressing fatty acid uptake has therapeutic effects in preclinical models of prostate cancer. Science Translational Medicine. 2019;**11**(478):eaau 5758. DOI: 10.1126/scitranslmed.aau5758

[15] Schroeder F, McIntosh AL, Martin GG, et al. Fatty acid binding

Protein-1 (FABP1) and the human FABP1 T94A variant: Roles in the endocannabinoid system and dyslipidemias. Lipids. 2016;**51**(6):655-676. DOI: 10.1007/s11745-016-4155-8

[16] Gajda AM, Storch J. Enterocyte fatty acid-binding proteins (FABPs): Different functions of liver and intestinal FABPs in the intestine. Prostaglandins, Leukotrienes, and Essential Fatty Acids. 2015;**93**:9-16. DOI: 10.1016/j.plefa.2014.10.001

[17] Yamamoto Y, Owada Y. Possible involvement of fatty acid binding proteins in psychiatric disorders. Anatomical Science International. 2021;**96**(3):333-342. DOI: 10.1007/s12565-020-00598-0

[18] Zeng J, Sauter ER, Li B. FABP4: A new player in obesity-associated breast cancer. Trends in Molecular Medicine. 2020;**26**(5):437-440. DOI: 10.1016/j.molmed.2020.03.004

[19] Uehara H, Kobayashi T, Matsumoto M, Watanabe S, Yoneda A, Bando Y. Adipose tissue: Critical contributor to the development of prostate cancer. The Journal of Medical Investigation. 2018;**65**(1.2):9-17. DOI: 10.2152/jmi.65.9

[20] Guaita-Esteruelas S, Gumà J, Masana L, Borràs J. The peritumoural adipose tissue microenvironment and cancer. The roles of fatty acid binding protein 4 and fatty acid binding protein 5. Molecular and Cellular Endocrinology. 2018;**462**(Pt B):107-118. DOI: 10.1016/j.mce.2017.02.002

[21] Zwicker BL, Agellon LB. Transport and biological activities of bile acids. The International Journal of Biochemistry & Cell Biology. 2013;**45**(7):1389-1398. DOI: 10.1016/j.biocel.2013.04.012

[22] Praslickova D, Torchia EC, Sugiyama MG, et al. The ileal lipid binding protein is required for efficient absorption and transport of bile acids in the distal portion of the murine small intestine. PLoS One. 2012;**7**(12):e50810. DOI: 10.1371/journal.pone.0050810

[23] Choi WS, Xu X, Goruk S, et al. FABP7 facilitates uptake of docosahexaenoic acid in glioblastoma neural stem-like cells. Nutrients. 2021;**13**(8):2664. DOI: 10.3390/nu13082664

[24] Abe M, Makino A, Murate M, et al. PMP2/FABP8 induces PI(4,5)P2-dependent transbilayer reorganization of sphingomyelin in the plasma membrane. Cell Reports. 2021;**37**(6):109935. DOI: 10.1016/j.celrep.2021.109935

[25] Moradi A, Forootan FS, Hosseini M, Pouresmaeili F. Leydig cells express the FABP9 in human testis. Human Antibodies. 2019;**27**(4):275-278. DOI: 10.3233/HAB-190382

[26] Hatzivassiliou G, Zhao F, Bauer DE, et al. ATP citrate lyase inhibition can suppress tumor cell growth. Cancer Cell. 2005;**8**(4):311-321. DOI: 10.1016/j.ccr.2005.09.008

[27] Zaidi N, Swinnen JV, Smans K. ATP-citrate lyase: A key player in cancer metabolism. Cancer Research. 2012;**72**(15):3709-3714. DOI: 10.1158/0008-5472.CAN-11-4112

[28] Hanai J, Doro N, Sasaki AT, et al. Inhibition of lung cancer growth: ATP citrate lyase knockdown and statin treatment leads to dual blockade of mitogen-activated protein kinase (MAPK) and phosphatidylinositol-3-kinase (PI3K)/AKT pathways. Journal of Cellular Physiology. 2012;**227**(4):1709-1720. DOI: 10.1002/jcp.22895

[29] Zong H, Zhang Y, You Y, Cai T, Wang Y. Decreased Warburg effect induced by ATP citrate lyase suppression inhibits tumor growth in pancreatic cancer [retracted in: Med Oncol. 2015 Oct; 32 (10):243]. Medical Oncology. 2015;**32**(3):85. DOI: 10.1007/s12032-015-0540-z

[30] Schug ZT, Gottlieb E. Cardiolipin acts as a mitochondrial signalling platform to launch apoptosis. Biochimica et Biophysica Acta. 2009;**1788**(10):2022-2031. DOI: 10.1016/j.bbamem.2009.05.004

[31] Kiebish MA, Han X, Cheng H, Chuang JH, Seyfried TN. Cardiolipin and electron transport chain abnormalities in mouse brain tumor mitochondria: Lipidomic evidence supporting the Warburg theory of cancer. Journal of Lipid Research. 2008;**49**(12):2545-2556. DOI: 10.1194/jlr.M800319-JLR200

[32] Schepers A, Clevers H. Wnt signaling, stem cells, and cancer of the gastrointestinal tract. Cold Spring Harbor Perspectives in Biology. 2012;**4**(4):a007989. Published 2012 Apr 1. DOI: 10.1101/cshperspect.a007989

[33] Anastas JN, Moon RT. WNT signalling pathways as therapeutic targets in cancer. Nature Reviews. Cancer. 2013;**13**(1):11-26. DOI: 10.1038/nrc3419

[34] Levental I, Grzybek M, Simons K. Greasing their way: Lipid modifications determine protein association with membrane rafts. Biochemistry. 2010;**49**(30):6305-6316. DOI: 10.1021/bi100882y

[35] Nakamura T. The roles of lipid mediators in type I hypersensitivity. Journal of Pharmacological Sciences. 2021;**147**(1):126-131. DOI: 10.1016/j.jphs.2021.06.001

[36] Pyne NJ, El Buri A, Adams DR, Pyne S. Sphingosine 1-phosphate and cancer. Advances in Biologial Regulation. 2018;**68**:97-106. DOI: 10.1016/j.jbior.2017.09.006

[37] Park JB, Lee CS, Jang JH, et al. Phospholipase signalling networks in cancer. Nature Reviews. Cancer. 2012;**12**(11):782-792. DOI: 10.1038/nrc3379

[38] Griner EM, Kazanietz MG. Protein kinase C and other diacylglycerol effectors in cancer. Nature Reviews. Cancer. 2007;**7**(4):281-294. DOI: 10.1038/nrc2110

[39] Vanhaesebroeck B, Stephens L, Hawkins P. PI3K signalling: The path to discovery and understanding. Nature Reviews. Molecular Cell Biology. 2012;**13**(3):195-203. Published 2012 Feb 23. DOI: 10.1038/nrm3290

[40] Sukocheva OA, Furuya H, Ng ML, et al. Sphingosine kinase and sphingosine-1-phosphate receptor signaling pathway in inflammatory gastrointestinal disease and cancers: A novel therapeutic target. Pharmacology & Therapeutics. 2020;**207**:107464. DOI: 10.1016/j.pharmthera.2019.107464

[41] De Craene B, Berx G. Regulatory networks defining EMT during cancer initiation and progression. Nature Reviews. Cancer. 2013;**13**(2):97-110. DOI: 10.1038/nrc3447

[42] Savagner P. The epithelial-mesenchymal transition (EMT) phenomenon. Annals of Oncology. 2010;**21**(Suppl. 7):vii89-vii92. DOI: 10.1093/annonc/mdq292

[43] Zhao W, Prijic S, Urban BC, et al. Candidate Antimetastasis drugs suppress the metastatic capacity of breast cancer

cells by reducing membrane fluidity. Cancer Research. 2016;**76**(7):2037-2049. DOI: 10.1158/0008-5472.CAN-15-1970

[44] Li T, Kang G, Wang T, Huang H. Tumor angiogenesis and anti-angiogenic gene therapy for cancer. Oncology Letters. 2018;**16**(1):687-702. DOI: 10.3892/ol.2018.8733

[45] Folkman J. Role of angiogenesis in tumor growth and metastasis. Seminars in Oncology. 2002;**29**(6 Suppl 16):15-18. DOI: 10.1053/sonc.2002.37263

[46] Kazlauskas A. Lysophosphatidic acid contributes to angiogenic homeostasis. Experimental Cell Research. 2015;**333**(2):166-170. DOI: 10.1016/j.yexcr.2014.11.012

[47] Mendelson K, Evans T, Hla T. Sphingosine 1-phosphate signalling. Development. 2014;**141**(1):5-9. DOI: 10.1242/dev.094805

[48] Teng MW, Galon J, Fridman WH, Smyth MJ. From mice to humans: Developments in cancer immunoediting. The Journal of Clinical Investigation. 2015;**125**(9):3338-3346. DOI: 10.1172/JCI80004

[49] Sung H, Ferlay J, Siegel RL, et al. Global cancer statistics 2020: GLOBOCAN estimates of incidence and mortality worldwide for 36 cancers in 185 countries. CA: A Cancer Journal for Clinicians. 2021;**71**(3):209-249. DOI: 10.3322/caac.21660

[50] Bray F, Laversanne M, Weiderpass E, Soerjomataram I. The ever-increasing importance of cancer as a leading cause of premature death worldwide. Cancer. 2021;**127**(16):3029-3030. DOI: 10.1002/cncr.33587

[51] Nakagawa H, Hayata Y, Kawamura S, Yamada T, Fujiwara N, Koike K. Lipid metabolic reprogramming in hepatocellular carcinoma. Cancers (Basel). 2018;**10**(11):447. DOI: 10.3390/cancers10110447

[52] Veigel D, Wagner R, Stübiger G, et al. Fatty acid synthase is a metabolic marker of cell proliferation rather than malignancy in ovarian cancer and its precursor cells. International Journal of Cancer. 2015;**136**(9):2078-2090. DOI: 10.1002/ijc.29261

[53] Pizer ES, Chrest FJ, DiGiuseppe JA, Han WF. Pharmacological inhibitors of mammalian fatty acid synthase suppress DNA replication and induce apoptosis in tumor cell lines. Cancer Research. 1998;**58**(20):4611-4615

Section 4

Lipid Metabolism, Autoimmune Diseases and Inflammation

Chapter 7

Lipid Metabolism and Associated Molecular Signaling Events in Autoimmune Disease

Mohan Vanditha, Sonu Das and Mathew John

Abstract

Lipid metabolism, when dysregulated paves the way to many autoimmune disease conditions. One such recently explored mechanism was that of Liver X receptor (LXR) signaling which acts as a molecular link between lipid metabolism and inflammation. LXR plays a critical role in coupling immune cell lipid homeostasis with systemic immune responses. In this chapter, we will discuss how an altered lipid metabolite environment causes inflammation signaling via LXR-mediated molecular events which could lead to autoimmune disease. In a hyperlipidemic environment, Interferon regulatory factor 3(IRF3) mediated downregulation of LXR signaling in innate immune cells leading to an inflammatory auto-immune response. Meanwhile, dendritic cell-mediated cytokine generation amidst LXR downregulation leads to the differentiation of autoreactive T cells and B cells, conferring an autoimmune response. Recent advances in the therapeutic management of autoimmune diseases target specific metabolic events as a strategy to limit inflammation and the autoimmune outcome. Novel treatment regimes in autoimmune diseases featuring lipid metabolic pathways are also discussed.

Keywords: LXR, hyperlipidemia, cholesterol overload, macrophage differentiation, oxysterols

1. Introduction

Autoimmune diseases (AIDs) represent a heterogeneous group of disorders that afflict specific target organs or multiple organ systems and are chronic conditions instigated by the loss of immunological tolerance to self-antigens [1]. It has also been noted that the overactive immune system that drives autoimmune diseases, presents metabolic abnormalities that provide therapeutic opportunities. Dysregulated lipid metabolism impacts the pathogenesis of AIDs such as rheumatoid arthritis (RA), Multiple sclerosis, and Systemic lupus erythematosus (SLE). Immunometabolism- the new emerging field has unveiled how bioenergetic and biosynthetic needs of healthy immune cells achieve rapid growth and perform effector functions upon challenge with pathogens [2]. Studies have shown that lipid-lowering drugs improve SLE symptomatically. Therefore, it is essential to study the molecular inflammatory signaling events associated with lipid metabolism in the context of AIDs [3]. This chapter comprehensively reviews Liver-X-Receptor (LXR) mediated regulation of inflammation and its downregulation in a hyperlipidemic environment, which is the highlight of this study.

Endogenous bioactive lipids play a decisive role in inflammatory processes and in triggering, coordinating, and confining immunity. In the case of excessive pro-inflammatory activity, these lipids contribute to the transition from acute to chronic inflammation. The major bioactive lipids involved in atherogenesis are eicosanoids, lysoglycerophospholipids, sphingolipids, and endocannabinoids. In-depth knowledge of the mechanism of action of these lipids in the pathogenesis of AIDs is required for the proper management of these diseases [4].

Atherogenic factors such as cholesterol oxidation products, malondialdehyde, and other aldehydes; trans-fatty acids; some saturated fatty acids (lauric, myristic, and possibly palmitic acids); and myristic acid plus cholesterol are involved in the regulation of innate and adaptive immune responses and lead to exacerbation of autoimmune diseases such as psoriasis, RA, and SLE.The phenotypes and functions of innate immune cells such as dendritic cells and macrophages are influenced by atherogenic risk factors, involved in lipid metabolism. Thus atherogenic risk factors have a major role in shaping the function of adaptive immune cells such as T and B cells. Microfluidics tools combined with biotechnological techniques and lipidomics have emphasized the role of different types of functional lipids and their derivatives in AIDs [5].

LXR signaling plays a critical role in coupling immune cell cholesterol homeostasis with systemic immune responses. This suggests that promoting reverse cholesterol transport via LXR signaling could have therapeutic utility in autoimmune diseases [6]. Cholesterol-lowering treatments such as a low-fat diet or statins were shown to be effective in ameliorating autoimmune symptoms [7].

2. The mechanism of LXR mediated autoimmunity

LXRs belong to the subfamily 1 of the nuclear hormone receptors super family (thyroid hormone receptor-like) that regulates cholesterol and lipid metabolism as well as inflammatory gene expression including nuclear factor-κB (NF-κB) and activator protein-1 (AP1). LXR deficiency and hypercholesterolemia lead to the accumulation of cholesterol in antigen presenting cells (APCs) including macrophages. The accumulation improves antigen presentation as well as T cell priming and the production of B cell activation factor (BAFF) and A proliferation-inducing ligand (APRIL) by APCs, all of which increase B cell differentiation and thus autoantibody production [6]. Thus, an inflammatory autoimmune response is established and leads to autoimmune disease pathogenesis.

Atypical increase of lipid species in plasma levels due to alteration in lipid metabolism sequentially stimulates innate immune cells like macrophages and dendritic cells through the recognition of the lipids via their receptors. A group of nuclear receptors is involved in metabolic balance, the sensing, and the export of intracellular lipid species. Among them, LXR induces cholesterol transporters on the cell surface that mediate the export of intracellular cholesterol. LXRs are critical regulators of cholesterol and fatty acid metabolism along with the regulation of inflammatory gene expression [8].

Antigen-presenting cells such as macrophages function to scavenge pathogens and apoptotic cells as well as to coordinate the inflammatory response to such stimuli through the production of cytokines and other mediators. The activation and inflammatory function of macrophages is modulated by several lipid species such as modified low-density lipoproteins (LDLs), fatty acids, and cholesterol crystals through LXR signaling. Gene expression studies in activated macrophages have revealed that LXRs inhibit genes involved in the innate immune response while simultaneously inducing those involved in lipogenesis. LXR acts as a molecular link between lipid metabolism and by antagonizing inflammatory gene expression and reducing inflammation [9].

2.1 LXR inhibition leads to autoimmunity in a hyperlipidemic environment

In a hyperlipidemic environment, macrophages form foam cells. Foam cells are cytoplasmic lipid droplets formed as a result of the accumulation of cholesterol esters, by virtue of uncontrolled uptake of oxidized low-density lipoprotein (ox-LDL), excessive cholesterol esterification and impaired cholesterol release [10]. Foam cell formation in macrophages has been shown to activate the NLR (Nod Like Receptor) family pyrin domain containing 3 (NLRP3)/inflammasome with the secretion of the proinflammatory cytokines, interleukin-1 β (IL-1β) and IL-18 and to promote the progression of atherogenesis [11]. Activation of foam cells can lead to the production of cytokines like IL-23, IL-6, and IL-27. IL 23 leads to tissue differentiation whereas IL-27 and IL-6 induce Tfh cell proliferation and differentiation. Tfh cells are newly identified CD4+ T helper subsets that mainly drive autoimmune germinal center reaction and autoantibody responses, which exacerbate autoimmune symptoms like an immune complex deposition. The hyperlipidemia-IL-27-Tfh cell axis is quintessential in the development of atherosclerosis-mediated SLE.

Dendritic cells, by virtue of the strength of antigen presentation and cytokine environment, govern CD4+ T cell activation and differentiation. It has been shown that hyperlipidemia and the nature of lipid species regulate antigen stimulating capacity and cytokine production of dendritic cells. Hyperlipidemic condition promotes the production of proinflammatory cytokines such as IL-1β, IL-6, and IL-27. When compared with control mice, mice subjected to a high-fat diet exhibited increased numbers of CD11b + dendritic cells, which secrete more IL-1β. Cholesterol accumulation in dendritic cells leads to autoimmune phenotypes such as immune complex deposition in kidneys and increased plasma dsDNA antibodies in mice [11].

Immunostimulatory lipid species like LDLs and saturated fatty acids can stimulate immune responses via lectin-like oxidized low-density lipoprotein receptor-1 (LOX1), cluster of differentiation 36 (CD36), toll-like receptors 2 and 4 (TLR2 and TLR4) downstream signaling. The uptake of free fatty acids including palmitic acid and oleic acid increases the production of IL-23 and IL-1β from bone-marrow-derived dendritic cells (DCs) [12]. Recent studies have demonstrated that LDLs and oxLDL stimulation of dendritic cells enhances IL-6 and IL-27 production in CD36 and TLR4/Myeloid differentiation primary response 88 (MyD88) dependent manner [13]. Dendritic cells become more sensitive to immunostimulatory lipid species via the upregulation of lipid receptors. Dendritic cells induce CD36 expression on their surface and promote IL-6 release by uptake of oxLDL. Increased expression of lipid receptor CD36 due to upregulated transcriptional activity of LXR is positively correlated with autoimmune phenotypes suggesting the mutual relation between a lipid receptor and autoimmunity [14]. Dendritic cells in atherogenic conditions exhibit higher expression of pattern-recognition receptors such as LOX-1, TLR2, and TLR4, all of which are known lipid receptors. Down-regulated LXR expression in hyperlipidemic conditions enhances NF-κB signaling and the consequent production of proinflammatory cytokines. Activation of LXR signaling by administrating LXR agonist to dendritic cells, reduced IL-27 production as well as the production of IL-23 and IL-12, all of which contribute to the differentiation of autoimmune Tfh and Th17cells.

Phagocytosis of pathogens, such as Gram-negative bacteria by antigen processing cells, elicits a marked immune response. Pathogens contain molecules such as lipopolysaccharide (LPS) that are recognized by the Toll family of receptors (TLR) such as TLR-3 and 4. These in turn activate interferon regulatory factor-3 (IRF3)-mediated inhibition of LXRs on their target promoters [15]. This could be a possible mechanism by which LXR downregulation leads to an inflammatory autoimmune response.

The successful clearance of invading pathogens depends on the neutrophil's efficient migration into the infected tissues. LXR activation impaired the neutrophil chemotactic response toward chemokines like C-X-C motif chemokine ligand 2 (CXCL2) in a concentration-dependent manner. LXR activation in neutrophils represses neutrophil migration genes [16]. LXR downregulation leads to neutrophil chemotaxis and over recruitment of neutrophils to the infected site. This leads to marked inflammation at the site.

The cholesterol sensing function of LXR in macrophages is likely to be important for the scavenger function of these cells. Large amounts of fatty acids and cholesterol will be accumulated in cells by the internalization of apoptotic cells and cellular debris. In this context, the role of LXR is to activate the cholesterol efflux pathway to protect the cell from lipid overload. LXR activation in cholesterol-loaded cells limits the production of inflammatory mediators. An inflammatory response is not apt when apoptotic cells are being scavenged. The absence of inflammation is a hallmark of apoptotic body clearance [17]. Decreased apoptotic clearance could promote an inflammatory autoimmune response. LXR agonist inhibits the expression of inflammatory responses including IL-6 and IL-1β [18, 19]. LXR expression in macrophages has a negative effect on inflammatory responses through the regulation of NF-κB signaling.

Accumulated cholesterol in innate immune cells enhances NLRP3 inflammasome activity. It promotes IL-1β and IL-18 secretion and GM-CSF receptor expression in an LXR-independent manner to elevate IL-12, IL-6, and IL-23 production by CD11b + dendritic cells. OxLDL plays a vital role in the activation of dendritic cells to promote IL-6 and IL-1β production, which enhances susceptibility to autoimmune diseases by regulating pathogenic autoimmune Th17 and Tfh cell differentiation [20, 21]. Th17 cells are one of the key players in the pathogenesis of autoimmunity [22].

2.2 LXR activation by oxysterols

LXRs are activated by the oxysterols- 24(*S*),25-epoxycholesterol, and 24(*S*)-hydroxycholesterol. It is well known that oxysterols suppress *de novo* cholesterol biosynthesis as well as cellular uptake of cholesterol [23]. LXR and RXR (retinoid X receptor), have an amino-terminal transcriptional activation domain (AF-1), a ligand-binding domain (LBD), a DNA binding domain (DBD), domains responsible for nuclear translocation and dimerization, and a transcriptional activation domain (AF-2) at the extreme carboxyl terminus [24]. When oxysterols, or synthetic pharmacological agonists, trigger activation of LXRs, they heterodimerize with RXR and bind to target gene promoters on LXR-responsive-elements (LXREs) [25, 26].

2.3 Effects of LXR activation

2.3.1 LXR activation

MODE OF ACTION 1: LXR activates an 'inducible degrader of the low-density lipoprotein' (IDOL) receptor. IDOL in turn reduces the expression of low-density lipoprotein receptor (LDLRs) on the cell surface, reducing cholesterol intake into the cells [27, 28].

MODE OF ACTION 2: Liver-X-receptor-Nuclear receptor corepressor-Silencing mediator of retinoic acid and thyroid hormone receptor (LXR-NCoR-SMRT) complex prevents transcription of inflammatory genes leading to transrepression of NF-kB. NF-κB is the central transcriptional regulator of the innate immune response. Many of the genes inhibited by LXR are established targets of NF-κB signaling. NF-κB promotes autoimmunity as well as inflammation by mediating the activation and differentiation of autoimmune and inflammatory T cells, such as Th17 cells. Efficient and properly controlled NFκB signaling is important for mediating normal immune homeostasis and function and for preventing autoimmunity [29]. Analysis of the inducible nitric oxide synthase (iNOS) and cyclooxygenase-2 (COX-2) gene promoters indicates that inhibition of these genes by LXR is accomplished through antagonism of NF-κB [30, 31].

MODE OF ACTION 3: LXR activates efferocytosis receptor- Mer proto-oncogene Tyrosine Kinase (MERTK), which is linked to the resolution of inflammation [18]. It is also found that MERTK can reciprocally regulate LXR expression identifying a potential feedback loop that may function to tip the balance from inflammation to resolution and tissue repair [32]. When activation of LXRs is inhibited, MERTK is downregulated which leads to decreased apoptotic cell clearance and leads to autoimmunity.

MODE OF ACTION 4: LXRs can induce the synthesis of long-chain polyunsaturated fatty acids (lcPUFAs) such as omega 3 fatty acids and mono-unsaturated fatty acids (MUFA). LXR upregulates Stearoyl CoA desaturase (SCD) and Fatty Acid Synthase (FASN) in the fatty acid biosynthesis pathway [33, 34]. The presence of lcPUFAs can decrease transactivation mediated by NF-kB of inflammatory genes, modifying histone acetylation in their regulatory regions [35]. lcPUFAs have been shown to increase the production of eicosanoids and selected pro-resolving lipid mediators [36]. Pro-resolving lipid mediators include lipoxin, resolvin, protectin, and maresin families, collectively called specialized pro-resolving mediators (SPM) [37]. Increased LXRs activity can also induce macrophage polarization toward a more pro-resolving phenotype (M2), directly upregulating the expression of MERTK and inflammation resolution.

MODE OF ACTION 5: LXRs upregulate the expression of sterol transporters such as the ATP binding cassette (ABC) family members ABCA1 and ABCG1, together with the transcription factors sterol regulatory element-binding protein 1c (SREBP1c) and carbohydrate-response element-binding protein (ChREBP). These genes in turn regulate critical lipogenic pathways, counteracting aberrant cellular overload [38]. Induction of the sterol transporters Abca1 and Abcg1, in turn, promotes cholesterol efflux to lipid-poor apolipoprotein A-I (ApoA-I) and high-density lipoprotein (HDL) acceptors. As a result, excess cholesterol from the periphery is shuttled by HDL to the liver for excretion through the reverse cholesterol transport pathway [39]. Thus, LXR activation counteracts sterol overload and resolves inflammation.

LXR deficiency or LXR inhibition leads to increased cholesterol in macrophages or dendritic cells. Cholesterol plays a key role in the regulation of immune responses through different mechanisms. Cholesterol is required for membrane synthesis during cell expansion and also it is a key constituent of lipid rafts. As a result, any changes in cholesterol content could regulate raft dependent signaling of

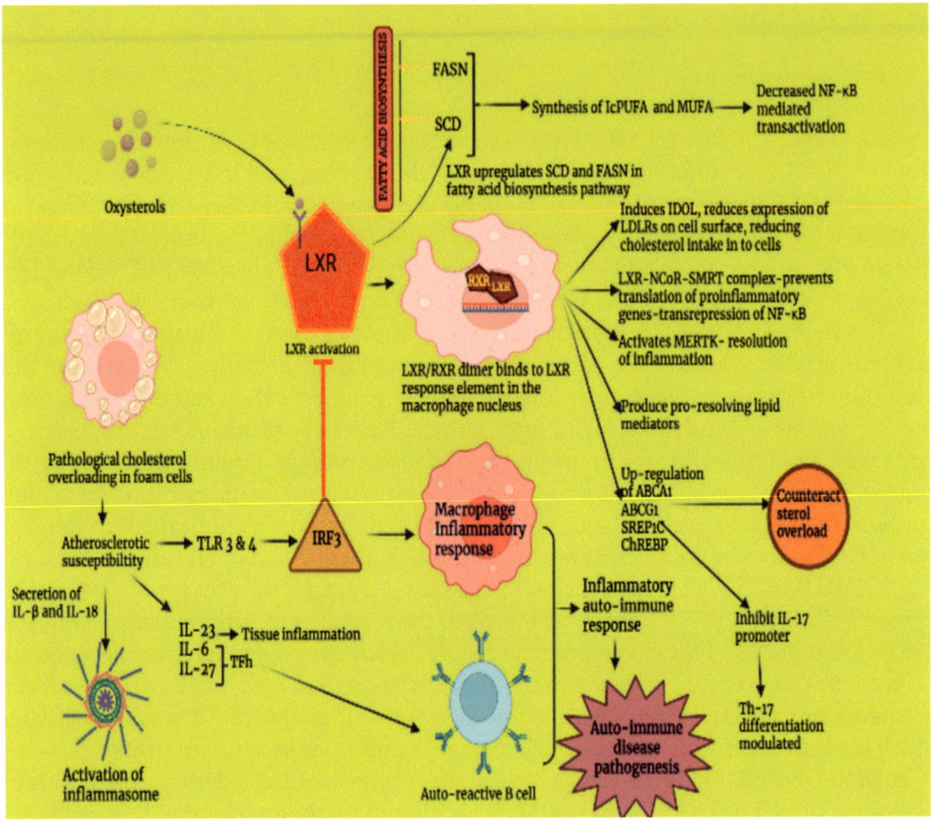

Figure 1.
LXR- a molecular link between lipid metabolism and inflammatory autoimmune response. FASN-Fatty acid synthase, LXR-Liver X Receptor, PUFA-Poly Unsaturated fatty acids, MUFA-mono Unsaturated fatty acids, SCD-Stearoyl-CoA Desaturase, RXR-Retinoid X Receptor, 1RF3- Interferon regulatory factor-3, IDOL-Inducible degrader of low-density lipoprotein, MERTK-Mer proto-oncogene tyrosine kinase, ABCA1-ATP binding cassette, SREBP-Sterol regulatory element-binding protein, and ChREBP-Carbohydrate response element-binding protein.

major immune pathways such as Toll-like receptors (TLRs), major histocompatibility complex (MHC), T cell receptors (TCRs) and B cell receptors (BCRs) [40]. Hyperlipidemia can activate TLR-3 and TLR-4 mediated IRF3 activation, which in turn inhibits LXR, leading to an inflammatory autoimmune response.

LXR activation and its effects are diagrammatically represented in **Figure 1**.

2.4 Adaptive immunity links autoimmunity with hyperlipidemia

Antigen presentation and cytokine production by innate immune cells are modulated by cellular lipid or cholesterol homeostasis and leads to direct activation of adaptive immune cells.

Naïve T-cells are differentiated into Th 17 cells by IL-6, transforming growth factor β (TGF β), IL-23, and IL-1 β stimulation, express RXR-related Orphan nuclear receptors (RORγt) and (RORα) and secrete IL-17A, IL-17Fand IL-22 to promote wound healing and eliminate extracellular pathogens via recruiting neutrophils. Th 17 cells lead to the autoimmune response either by modulating tissue inflammation or by IL-17 mediated autoantibody production. Tfh cells are differentiated by IL-6, IL-12,

IL-21, and IL-27 stimulation express B cell lymphoma 6 (Bcl6) and Achaete-scute complex homolog 2 (Ascl2), and secrete IL-21 [20, 41]. Proper activation of Th 17 and Tfh cells protect the body from infection but an uncontrolled generation of the cells can also contribute to the pathogenesis of autoimmune diseases.

B cells are responsible for the generation of pathogenic autoantibodies, thus intense research is carried out to study the function of autoreactive B cells and how they are triggered in autoimmune diseases. IL- 17 produced by Th17 cells is required for autoreactive B cell production and germinal center reactions [42, 43]. Furthermore, IL- 21, IFNγ, and IL-4 secreted by Tfh cells are required for class switching of IgG2a/c and IgG1, respectively, during Tcell- B cell interaction. IL-27 is sufficient to induce an increase in Tfh cells and germinal center reactions. Analysis of plasma from healthy controls and hypercholesterolemia patients showed that IL-27, but not IL-6, is increased in the patients with hypercholesterolemia and that IL-27 is associated with increased immunoglobulin G (IgG) in the circulation.

Tfh cells express Bcl 6 and secrete IL-21 which provides crucial help for B cells to induce class switching, affinity maturation, and differentiation into plasma cells through germinal center reactions [44]. A novel function of LXRs as modulators of

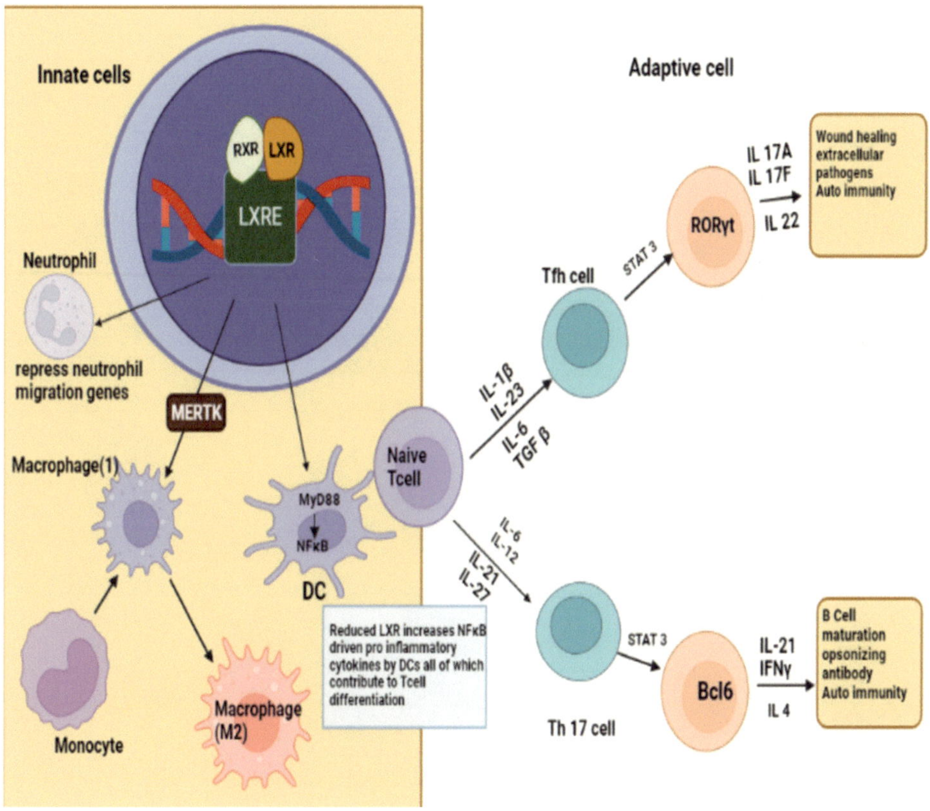

Figure 2.
LXR mediated immune cell differentiation and cytokine expression. LXR-Liver X Receptor, RXR-Retinoid X Receptor, LXRE-LXR responsive elements, MERTK- Mer proto-oncogene tyrosine kinase, DC-dendritic cells, RXR related orphan nuclear receptor-gamma, STAT 3-Signal transducer and activator of transcription 3, and BCL-B cell lymphoma.

B cell activation and in the control of Ig E synthesis suggests a beneficial function of LXRs in allergic therapy [45]. LXR downregulation leads to activation of Tfh cell that promotes B cell activation which results in an overall surge of autoantibodies regardless of immunoglobulin subclass. The major roles of Bcl6 in the development and effector function of mature T cell subsets suggest that Bcl6, besides, being a regulator of germinal center reactions, is also an important regulator of T cell dependant inflammatory, autoimmune and memory response in the periphery [46]. Oxysterol directly acts as an RXR-related Orphan nuclear receptor γt (RORγt) agonist to promote Th17 cell differentiation [47, 48].

LXR-mediated immune cell differentiation and cytokine expression are represented as shown in **Figure 2**.

3. Mechanism of LXR down-regulation leading to SLE clinical manifestation

Upregulated atherosclerotic conditions can downregulate LXR signaling, thereby downregulating the expression of various genes involved in the resolution of inflammation and autoimmune response.

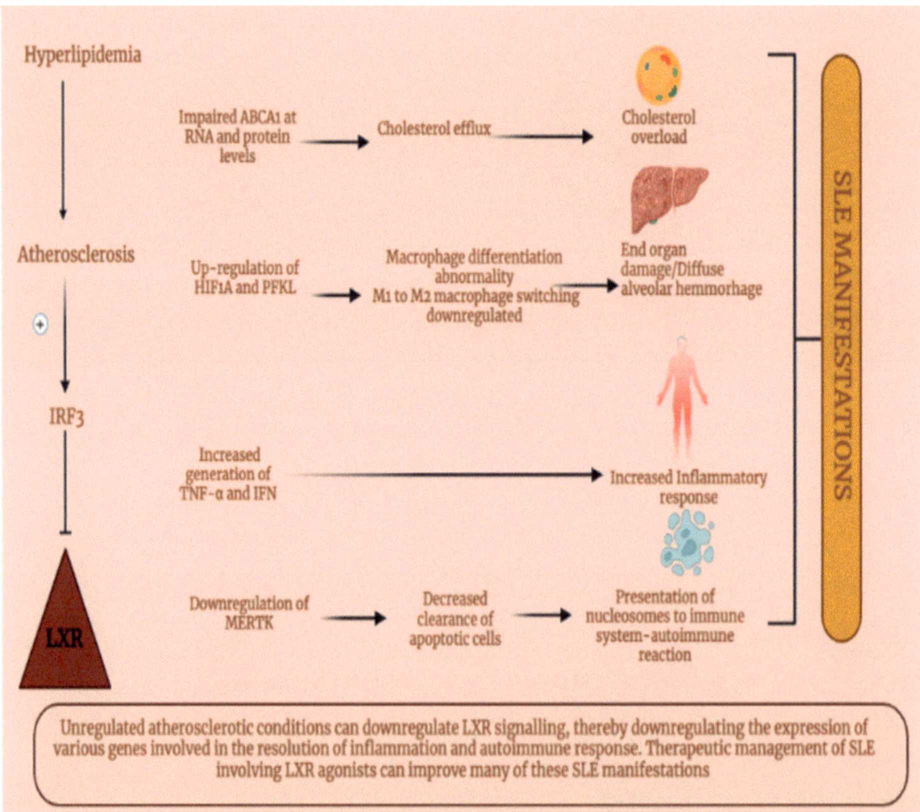

Figure 3.
Downregulated LXR mediated clinical manifestations in systemic lupus erythematosus. LXR-Liver X Receptor, 1RF3-Interferon regulatory factor-3 MERTK- Mer proto-oncogene tyrosine kinase, ABCA1- ATP binding cassette, TNF-Tumor necrosis factor, IFN- interferons, HIF1A- Hypoxia-inducible factor-1-alpha, 6-phosphofructokinase, liver type.

Downregulation of LXR signaling leads to impaired ABCA1 at RNA and protein levels which causes cholesterol efflux from cell to periphery. This results in cholesterol overload. Upregulation of Hypoxia Inducible Factor 1A (HIF1A) and 6-phosphofructokinase, liver type (PFKL) genes leads to abnormal macrophage differentiation. Switching from M1 TO M2 phenotype is down-regulated which triggers diffuse alveolar hemorrhage or end-organ damage in the liver and other vital organs [49].

Downregulation of LXR leads to increased generation of pro-inflammatory cytokines like Tumor necrosis factor (TNF) α and Interferon γ which causes an increased inflammatory response. Downregulation of MERTK genes leads to decreased efferocytosis, ie decreased clearance of apoptotic cells that result in autoimmune reactions via nucleosome presentation [50].

Therapeutic management of SLE involving LXR agonists can improve many of these SLE manifestations. A diagram outlining the mechanism of hyperlipidemia-mediated LXR downregulation in SLE is shown in **Figure 3**.

4. Mechanism of LXR down-regulation leading to RA clinical manifestations

Hyperlipidemia-mediated downregulation of LXR signaling can lead to many of the pathophysiological symptoms observed in rheumatoid arthritis. This LXR inhibition in turn could activate the production of pro-inflammatory cytokines such as C-X-C motif chemokine ligand 10 (CXCL10). Fibroblast-like synoviocyte (FLS) invasion, one of the major symptoms of RA, leading to joint damage could be due to an increased expression of CXCL10 [51]. Likewise, FLS invasion could also be brought about by the expression of another set of pro-inflammatory cytokines like IL-1 and IL-6, leading to matrix metalloprotease 2 (MMP-2) expression. Cartilage destruction and joint damage, another debilitating manifestation of RA is also brought about by MMP-3 expression. Moreover, LXR-mediated differentiation of Th cells could also lead to Synovial tissue hyperplasia causing tissue invasion and cartilage destruction [52].

LXR agonists as a therapeutic option for rheumatoid arthritis can suppress pro-inflammatory cytokines such as IL-1 and ILN-6 as well as modulate Th17 differentiation, thereby alleviating symptoms in rheumatoid arthritis.

5. Treatment

Therapeutic target 1: It is well known that ω-3 PUFAs have anti-inflammatory properties and their presence in nutrition contributes to the prevention of many inflammatory diseases, independently of the mechanism of action [53]. The administration or inclusion of ω-3 PUFAs in the human diet appears as the most natural way to reduce AIDs symptoms. In particular, the availability of ω-3 PUFAs in the human diet could dramatically change their benefits (**Figure 4**) [54].

Therapeutic target 2: Statins, or 3-hydroxy-3-methylglutaryl coenzyme A (HMG-CoA) reductase inhibitors, have emerged as the leading therapeutic regimen for treating hypercholesterolemia. Statins can accomplish partial inhibition of an upstream common denominator of multiple regulatory signaling networks that control the immune system. Thus, statins may be a good choice for ameliorating AID symptoms.

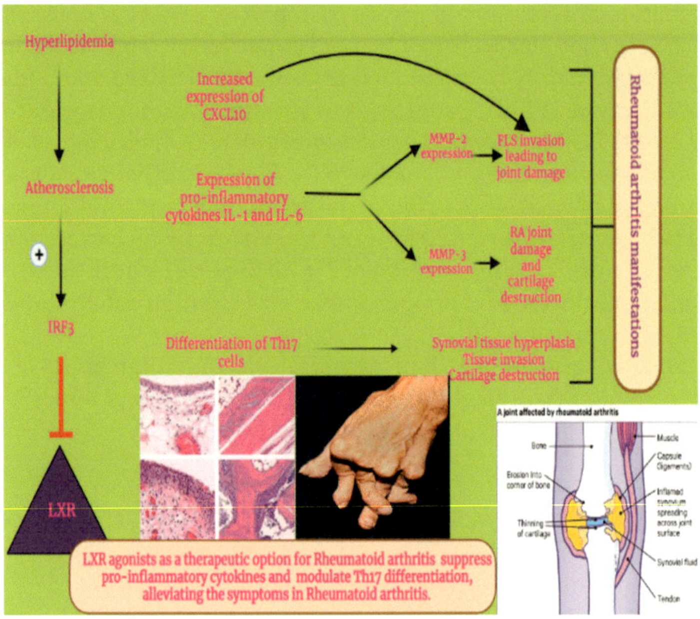

Figure 4.
Downregulated LXR mediated clinical manifestations in rheumatoid arthritis. LXR-Liver X Receptor, 1RF3-Interferon regulatory factor-3, CXCL10- C-X-C motif chemokine ligand 10, Matrix metalloproteinases, and FLS-Fibroblast like synoviocytes.

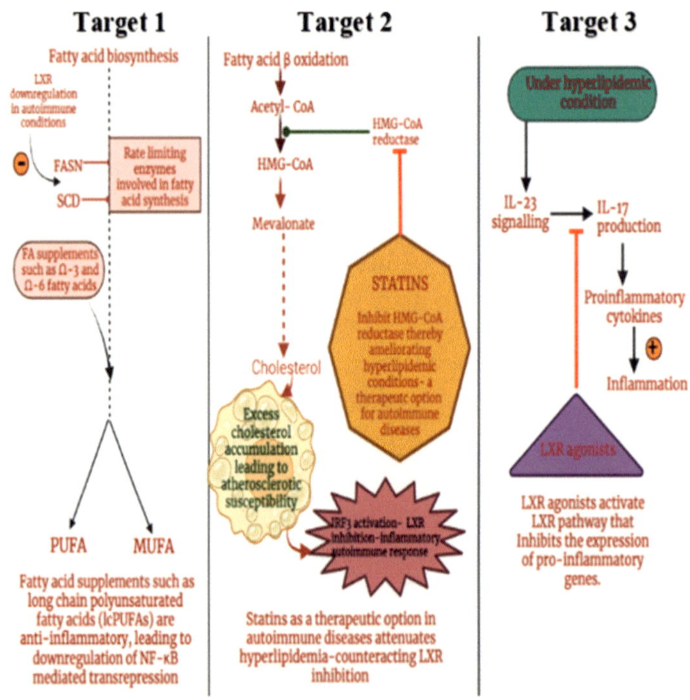

Figure 5.
Therapeutic management of auto-immune diseases. FASN-Fatty acid synthase, LXR-Liver X Receptor, PUFA-Poly Unsaturated fatty acids, MUFA-mono Unsaturated fatty acids, SCD-Stearoyl-CoA Desaturase, 1RF3-Interferon regulatory factor-3, HMG-CoA- 3-Hydroxy-3-methylglutaryl-coenzyme a reductase.

Therapeutic target 3: LXR agonists activate the LXR pathway inhibiting the expression of pro-inflammatory genes. They also increase the flux of cholesterol into the liver, where it is metabolized.

The therapeutic management of Auto-immune diseases is summarized as shown in **Figure 5**.

All the illustrations are original and created by the authors using BioRender, an online web application used to create scientific figures and diagrams.

6. Conclusion

In the present review, the authors analyzed the different aspects of lipid metabolism which contribute to autoimmune disease via inflammatory signaling pathways. Reducing the hyperlipidemic environment could alleviate the pathophysiological complications of auto-immune diseases, by modulating the Liver-X-Receptor (LXR) signaling. LXR agonists along with fatty acid supplementation and statins are promising therapeutic targets for efficient clinical management of auto-immune diseases such as rheumatoid arthritis and systemic lupus erythematosus.

Acknowledgements

The authors greatly acknowledge the support of the Jubilee Centre for Medical Research. The authors are also thankful to Dr.D.M. Vasudevan and Dr.P.R. Varghese for their encouragement.

Conflict of interest

The authors declare no conflict of interest.

Author details

Mohan Vanditha, Sonu Das and Mathew John*
Jubilee Centre for Medical Research, Jubilee Mission Medical College and Research Institute, Thrissur, Kerala, India

*Address all correspondence to: mathewjohn@jmmc.ac.in

References

[1] Anaya JM. Common mechanisms of autoimmune diseases (the autoimmune tautology). Autoimmunity Reviews. 2012;**11**(11):781-784

[2] Pålsson-McDermott EM, O'Neill LAJ. Targeting immunometabolism as an anti-inflammatory strategy. Cell Research. 2020;**30**(4):300-314

[3] Ryu H, Kim J, Kim D, Lee JE, Chung Y. Cellular and molecular links between autoimmunity and lipid metabolism. Molecules and Cells. 2019;**42**(11):747-754

[4] Chiurchiù V, Leuti A, Maccarrone M. Bioactive lipids and chronic inflammation: Managing the fire within. Frontiers in Immunology. 2018;**29**:38

[5] Marques MP, Boyd AS, Polizzi K, Szita N. Microfluidic devices towards personalized health and wellbeing. Journal of Chemical Technology and Biotechnology. 2019;**94**(8):2412-2415

[6] Ito A, Hong C, Oka K, Salazar JV, Diehl C, Witztum JL, et al. Cholesterol accumulation in CD11c+ immune cells is a causal and targetable factor in autoimmune disease. Immunity. 2016;**45**(6):1311-1326

[7] Youssef S, Stüve O, Patarroyo JC, Ruiz PJ, Radosevich JL, Hur EM, et al. The HMG-CoA reductase inhibitor, atorvastatin, promotes a Th2 bias and reverses paralysis in central nervous system autoimmune disease. Nature. 2002;**420**(6911):78-84

[8] Kiss M, Czimmerer Z, Nagy L. The role of lipid-activated nuclear receptors in shaping macrophage and dendritic cell function: From physiology to pathology. The Journal of Allergy and Clinical Immunology. 2013;**132**(2):264-286

[9] Joseph SB, Castrillo A, Laffitte BA, Mangelsdorf DJ, Tontonoz P. Reciprocal regulation of inflammation and lipid metabolism by liver X receptors. Nature Medicine. 2003;**9**(2):213-219

[10] Yu XH, Fu YC, Zhang DW, Yin K, Tang CK. Foam cells in atherosclerosis. Clinica Chimica Acta. 2013;**424**:245-252

[11] Reynolds CM, McGillicuddy FC, Harford KA, Finucane OM, Mills KHG, Roche HM. Dietary saturated fatty acids prime the NLRP3 inflammasome via TLR4 in dendritic cells—implications for diet-induced insulin resistance. Molecular Nutrition & Food Research. 2012;**56**(8):1212-1222

[12] Stelzner K, Herbert D, Popkova Y, Lorz A, Schiller J, Gericke M, et al. Free fatty acids sensitize dendritic cells to amplify TH1/TH17-immune responses. European Journal of Immunology. 2016;**46**(8):2043-2053

[13] Lim H, Kim YU, Sun H, Lee JH, Reynolds JM, Hanabuchi S, et al. Proatherogenic conditions promote autoimmune T Helper 17 cell responses in vivo. Immunity. 2014;**40**(1):153-165

[14] Nickel T, Schmauss D, Hanssen H, Sicic Z, Krebs B, Jankl S, et al. oxLDL uptake by dendritic cells induces upregulation of scavenger-receptors, maturation and differentiation. Atherosclerosis. 2009;**205**(2):442-450

[15] Castrillo A, Joseph SB, Vaidya SA, Haberland M, Fogelman AM, Cheng G, et al. Crosstalk between LXR and Toll-like receptor signaling mediates bacterial and viral antagonism of cholesterol metabolism. Molecular Cell. 2003;**12**(4):805-816

[16] Souto FO, Castanheira FVS, Trevelin SC, Lima BHF, Cebinelli GCM, Turato WM, et al. Liver X receptor activation impairs neutrophil functions and aggravates sepsis. The Journal of Infectious Diseases. 2020;**221**(9):1542-1553

[17] Bilotta MT, Petillo S, Santoni A, Cippitelli M. Liver X receptors: Regulators of cholesterol metabolism, inflammation, autoimmunity, and cancer. Frontiers in Immunology. 2020;**11**:584303

[18] Thomas DG, Doran AC, Fotakis P, Westerterp M, Antonson P, Jiang H, et al. LXR suppresses inflammatory gene expression and neutrophil migration through cis-repression and cholesterol efflux. Cell Reports. 2018;**25**(13):3774-3785

[19] Pourcet B, Gage MC, León TE, Waddington KE, Pello OM, Steffensen KR, et al. The nuclear receptor LXR modulates interleukin-18 levels in macrophages through multiple mechanisms. Scientific Reports. 6 May 2016;**6**:25481

[20] Batten M, Ramamoorthi N, Kljavin NM, Ma CS, Cox JH, Dengler HS, et al. IL-27 supports germinal center function by enhancing IL-21 production and the function of T follicular helper cells. The Journal of Experimental Medicine. 2010;**207**(13):2895-2906

[21] Canavan M, McCarthy C, Larbi NB, Dowling JK, Collins L, O'Sullivan F, et al. Activation of liver X receptor suppresses the production of the IL-12 family of cytokines by blocking nuclear translocation of NF-κBp50. Innate Immunity. 2014;**20**(7):675-687

[22] Duewell P, Kono H, Rayner KJ, Sirois CM, Vladimer G, Bauernfeind FG, et al. NLRP3 inflammasomes are required for atherogenesis and activated by cholesterol crystals. Nature. 2010;**464**(7293):1357-1361

[23] Lehmann JM, Kliewer SA, Moore LB, Smith-Oliver TA, Oliver BB, Su JL, et al. Activation of the nuclear receptor LXR by Oxysterols defines a new hormone response pathway*. The Journal of Biological Chemistry. 1997;**272**(6):3137-3140

[24] Edwards PA, Kast HR, Anisfeld AM. BAREing it all: The adoption of LXR and FXR and their roles in lipid homeostasis. Journal of Lipid Research. 2002;**43**(1):2-12

[25] Pehkonen P, Welter-Stahl L, Diwo J, Ryynänen J, Wienecke-Baldacchino A, Heikkinen S, et al. Genome-wide landscape of liver X receptor chromatin binding and gene regulation in human macrophages. BMC Genomics. 2012;**13**(1):50

[26] Jakobsson T, Treuter E, Gustafsson JÅ, Steffensen KR. Liver X receptor biology and pharmacology: New pathways, challenges and opportunities. Trends in Pharmacological Sciences. 2012;**33**(7):394-404

[27] Hong C, Duit S, Jalonen P, Out R, Scheer L, Sorrentino V, et al. The E3 ubiquitin ligase idol induces the degradation of the low density lipoprotein receptor family members VLDLR and ApoER2. The Journal of Biological Chemistry. 2010;**285**(26):19720-19726

[28] Tontonoz P. Transcriptional and posttranscriptional control of cholesterol homeostasis by liver X receptors. Cold Spring Harbor Symposia on Quantitative Biology. 2011;**76**:129-137

[29] Sun SC, Chang JH, Jin J. Regulation of nuclear factor-κB in autoimmunity. Trends in Immunology. 2013;**34**(6): 282-289

[30] Zhu M, Fu Y. The complicated role of NF-κB in T-cell selection. Cellular & Molecular Immunology. 2010;**7**(2):89-93

[31] Aschenbrenner K, D'Cruz LM, Vollmann EH, Hinterberger M, Emmerich J, Swee LK, et al. Selection of Foxp3+ regulatory T cells specific for self antigen expressed and presented by Aire+ medullary thymic epithelial cells. Nature Immunology. 2007;**8**(4):351-358

[32] Choi JY, Seo JY, Yoon YS, Lee YJ, Kim HS, Kang JL. Mer signaling increases the abundance of the transcription factor LXR to promote the resolution of acute sterile inflammation. Science Signaling. 2015;**8**(365):ra21

[33] Chu K, Miyazaki M, Man WC, Ntambi JM. Stearoyl-Coenzyme A Desaturase 1 deficiency protects against hypertriglyceridemia and increases plasma high-density lipoprotein cholesterol induced by liver X receptor activation. Molecular and Cellular Biology. 2006;**26**(18):6786-6798

[34] Kim HJ, Miyazaki M, Ntambi JM. Dietary cholesterol opposes PUFA-mediated repression of the stearoyl-CoA desaturase-1 gene by SREBP-1 independent mechanism. Journal of Lipid Research. 2002;**43**(10):1750-1757

[35] Li P, Spann NJ, Kaikkonen M, Lu M, Oh D, Fox J, et al. NCOR repression of LXRs restricts macrophage biosynthesis of insulin-sensitizing omega 3 fatty acids. Cell. 2013;**155**(1):200-214

[36] Serhan CN, Chiang N, Van Dyke TE. Resolving inflammation: Dual anti-inflammatory and pro-resolution lipid mediators. Nature Reviews. Immunology. 2008;**8**(5):349-361

[37] Serhan CN. Pro-resolving lipid mediators are leads for resolution physiology. Nature. 2014;**510**(7503):92-101

[38] Horton JD, Goldstein JL, Brown MS. SREBPs: Activators of the complete program of cholesterol and fatty acid synthesis in the liver. The Journal of Clinical Investigation. 2002;**109**(9):1125-1131

[39] Murphy AJ, Akhtari M, Tolani S, Pagler T, Bijl N, Kuo CL, et al. ApoE regulates hematopoietic stem cell proliferation, monocytosis, and monocyte accumulation in atherosclerotic lesions in mice. The Journal of Clinical Investigation. 2011;**121**(10):4138-4149

[40] Wang S, Guang YS, Quan PD, Ping ZS. HDL and ApoA-I inhibit antigen presentation-mediated T cell activation by disrupting lipid rafts in antigen presenting cells. Atherosclerosis. 2012;**225**(1):105-114

[41] Nurieva RI, Chung Y, Hwang D, Yang XO, Kang HS, Ma L, et al. Generation of T follicular helper cells is mediated by Interleukin-21 but Independent of T Helper 1, 2, or 17 cell lineages. Immunity. 2008;**29**(1):138-149

[42] Hsu HC, Yang P, Wang J, Wu Q, Myers R, Chen J, et al. Interleukin 17–producing T helper cells and interleukin 17 orchestrate autoreactive germinal center development in autoimmune BXD2 mice. Nature Immunology. 2008;**9**(2):166-175

[43] Mitsdoerffer M, Lee Y, Jäger A, Kim HJ, Korn T, Kolls JK, et al. Proinflammatory T helper type 17 cells are effective B-cell helpers. Proceedings of the National Academy of Sciences. 2010;**107**(32):14292-14297

[44] Wilson CS, Elizer SK, Marshall AF, Stocks BT, Moore DJ. Regulation of

B lymphocyte responses to Toll-like receptor ligand binding during diabetes prevention in non-obese diabetic (NOD) mice. Journal of Diabetes. 2016;**8**(1):120-131

[45] Heine G, Dahten A, Hilt K, Ernst D, Milovanovic M, Hartmann B, et al. Liver X receptors control IgE expression in B cells. Journal of Immunology. 2009;**182**(9):5276-5282

[46] Bunting KL, Melnick AM. New effector functions and regulatory mechanisms of BCL6 in normal and malignant lymphocytes. Current Opinion in Immunology. 2013;**25**(3):339-346

[47] Santori FR, Huang P, van de Pavert SA, Douglass EF, Leaver DJ, Haubrich BA, et al. Identification of natural RORγ ligands that regulate the development of lymphoid cells. Cell Metabolism. 2015;**21**(2):286-298

[48] Soroosh P, Wu J, Xue X, Song J, Sutton SW, Sablad M, et al. Oxysterols are agonist ligands of RORγt and drive Th17 cell differentiation. Proceedings of the National Academy of Sciences. 2014;**111**(33):12163-12168

[49] Ahamada MM, Jia Y, Wu X. Macrophage polarization and plasticity in systemic lupus erythematosus. Frontiers in Immunology. 2021;**12**:734008

[50] Xu N, Li J, Gao Y, Zhou N, Ma Q, Wu M, et al. Apoptotic cell-mimicking gold nanocages loaded with LXR agonist for attenuating the progression of murine systemic lupus erythematosus. Biomaterials. 2019;**197**:380-392

[51] Laragione T, Gulko PS. Liver X receptor regulates rheumatoid arthritis fibroblast-like synoviocyte invasiveness, matrix metalloproteinase 2 Activation, Interleukin-6 and CXCL10. Molecular Medicine. 2012;**18**(6):1009-1017

[52] Rose BJ, Kooyman DL. A tale of two joints: The role of matrix metalloproteases in cartilage biology. Disease Markers. 2016;**2016**:4895050

[53] Cas MD, Roda G, Li F, Secundo F. Functional lipids in autoimmune inflammatory diseases. International Journal of Molecular Sciences. 2020;**21**(9):3074

[54] Joffre C, Rey C, Layé S. N-3 polyunsaturated fatty acids and the resolution of neuroinflammation. Frontiers in Pharmacology. 2019;**10**:1022

Chapter 8

Inflammatory Diseases and the Role of n-7 Unsaturated Fatty Acids as Functional Lipids

Akio Nakamura, Hikari Nakamura and Ritsuko Kawaharada

Abstract

With the increasing childbearing age, the number of mothers with diabetes and gestational diabetes is escalating. Maternal hyperglycemia creates an intrauterine hyperglycemic environment via the placenta, which causes signaling abnormalities in various fetal organs due to excessive glycation. This is associated with future disease development in the child. We have shown that insulin signaling defects are induced in fetal cardiomyoblasts using a rat gestational diabetes mellitus model and cellular models. Furthermore, we reported that maternal intake of eicosapentaenoic acid (EPA), an n-3 unsaturated fatty acid, during pregnancy can ameliorate this signaling defect. However, EPA has anti-coagulant effects, and the pollution of marine fish oil, the source for EPA supplements, raises concerns about active intake by pregnant women. Recently, palmitoleic acid, an n-7 unsaturated fatty acid, garnered attention as a candidate functional lipid alternative to EPA because it has been reported to have anti-obesity, lipid metabolism improvement, and cardioprotective effects similar to those of EPA. Palmitoleic acid has cis and trans structural isomers, which differ in their food intake route and metabolism in humans. This article introduces recent findings on the biological functions of palmitoleic acid in lifestyle-related diseases and cardiovascular diseases, ranging from basic research to clinical studies.

Keywords: dairy products, eicosapentaenoic acid, gestational diabetes mellitus, heart disease, hyperglycemia, intrauterine hyperglycemic environment, lipid metabolism, palmitic acid, palmitoleic acid, trans fatty acid

1. Introduction

Prenatal nutrition has a significant impact on the long-term health of the unborn child. In particular, the cardiovascular epidemiological studies by Barker et al. have shown that the intrauterine environment during pregnancy is closely related to the development of future lifestyle-related diseases [1–5]. Gluckman and Hanson proposed the Developmental Origins of Health and Disease hypothesis, which states that predisposition to lifestyle-related diseases is shaped by gene-environment interactions during fertilization, embryonic development, fetal life, and infancy, and that mismatches with the fetal environment after birth lead to the development of diabetes and hypertension [6, 7]. Many of these studies have suggested that a low-nutrition environment in the womb during the fetal period increases the likelihood that the child will later suffer from lifestyle-related diseases, such as obesity, diabetes, and hypertension in the future [8–10].

In the post-World War II period, economically developed countries entered an era of global food satiation. Economic growth also changed people's lifestyles. These social conditions have been accompanied by an increase in late marriages, an increase in maternal obesity, an increase in maternal age, an increase in the incidence of gestational diabetes mellitus (GDM), and an increase in pregnant women with type 2 diabetes. The nutritional environment of pregnant women has increasingly become overnourished rather than undernourished [11–13]. The fetus of a pregnant woman with such an abnormal glucose metabolism is exposed to an intrauterine hyperglycemic environment, through the umbilical cord, due to maternal hyperglycemia. As a result, neonatal complications, such as gigantism, hypoxia, respiratory disorders, congenital malformations, and myocardial hypertrophy, have been reported in children born to diabetic pregnant women [14–18]. However, the molecular mechanisms by which this intrauterine hyperglycemic environment during the fetal period contributes to the future health and disease development of the child after birth are not well understood. We are conducting basic research using animal and cellular models to elucidate the underlying molecular mechanisms and to search for foods that show diabetes-preventive effects in pregnant women. In this chapter, we present recent findings on functional lipids that may prevent cardiac disease in children born to diabetic mothers.

2. Fetal heart in an intrauterine hyperglycemic environment

We created a GDM model rat by administering streptozotocin into the tail vein of Wistar rats on gestation day 2. To determine the effects of a high-fat diet during pregnancy on the infants, we fed the GDM model rats a high-fat lard diet containing saturated fatty acids and a control diet [19]. The stillbirth rate of GDM model rats fed a high-fat lard diet was significantly higher than that of GDM model rats fed a control diet [19]. Palmitic acid loading in lard has been reported to cause inflammation and cardiomyocyte dysfunction in animal- and cellular-level experiments [20–25]. The GDM rat study suggested that not only was the fetus exposed to intrauterine hyperglycemia, but that the mother's intake of palmitic acid from a high-fat lard diet during gestation may have further impaired cardiac function in the offspring [19].

Based on reports that fish oil has a positive effect on cardiovascular disease [26, 27], we fed GDM model rats a high-fat fish oil diet, a high-fat lard diet, and a control diet and analyzed the nutrient signals from the hearts of the infants. We found that the lard-fed GDM model rat offspring had decreased phosphorylation levels of Akt, which is important for sugar uptake [28]. However, when mothers were fed fish oil during pregnancy, the Akt phosphorylation level was increased, and insulin signaling was improved in the hearts of GDM model rat offspring [28]. The biological functional components in fish oil may have ameliorated the signaling impairment caused by hyperglycemia and palmitic acid loading.

We isolated primary cardiomyoblasts from the hearts of rat infants and established a cellular model of a hyperglycemic environment by elevating the glucose concentration in the medium to address two questions: Why was insulin signaling abnormal in the hearts of GDM model rat offspring, and what components of fish oil ameliorated this signaling abnormality? We focused on eicosapentaenoic acid (C20:5n-3: EPA), which is abundant in fish oil and has known cardiovascular protective effects [29–33]. We orally administered EPA to GDM model rats during gestation and analyzed neonatal primary cardiomyoblasts isolated from the offspring. The results showed that insulin signaling was inhibited in primary cardiomyoblasts of GDM model rat offspring and that insulin resistance had been induced in these rat infants [34].

However, insulin resistance was improved in the hearts of infants born to GDM model rat mothers that were orally administered EPA [34]. Furthermore, in primary cardiomyocytes from neonatal rats exposed to an intrauterine hyperglycemic environment, intracellular reactive oxygen species (ROS) were chronically elevated, and excess advanced glycation end-products (AGEs) were observed, as compared with normal cells. AGE formation has recently attracted attention as a cause of aging and disease [35–40]. Furthermore, AGEs increase the expression of receptors for AGEs (RAGEs) and induce AGE-RAGE signaling. This signaling also increases the expression of genes encoding various inflammatory cytokines (IL-6, TNFα, NF-κB) via phosphorylation of Jun amino-terminal kinase (JNK).

These results indicated that the fetuses of GDM model rats are exposed to a hyperglycemic environment in utero, resulting in chronic inflammation by accumulating AGEs. However, feeding EPA to pregnant GDM model rats suppressed the production of AGEs and intracellular ROS in the offspring, resulting in the amelioration of signaling abnormalities [34]. Therefore, it is highly likely that EPA was responsible for the improvement of cardiac insulin signaling. In that regard, a summary is shown in **Figure 1**.

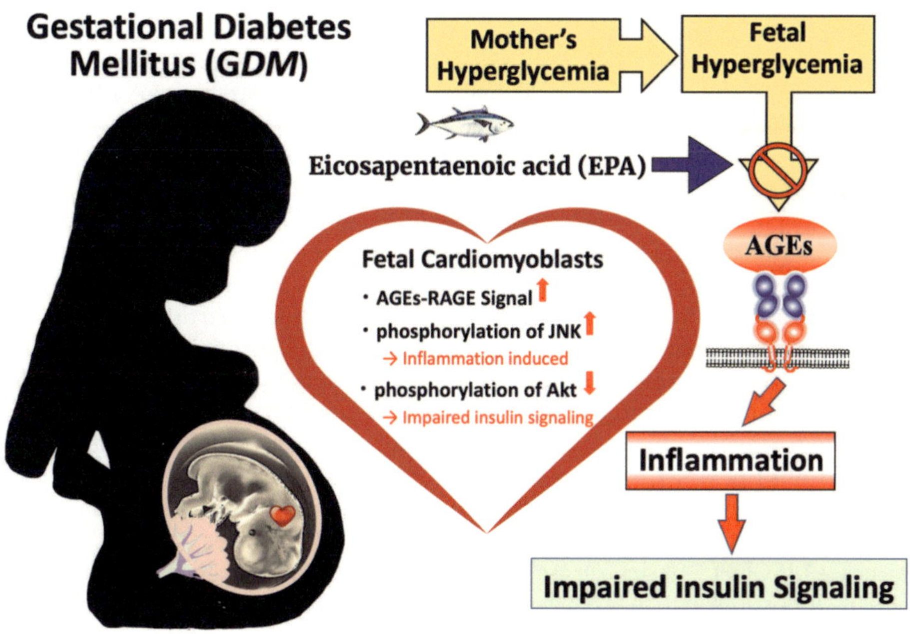

Figure 1.
Intrauterine hyperglycemia environment causes impaired signaling in the fetal heart. In gestational diabetes mellitus, maternal hyperglycemia creates a hyperglycemic intrauterine environment via the placenta. In this environment, advanced glycation end-products (AGEs) of proteins accumulate in fetal cardiomyoblasts, triggering inflammatory signals and impairing nutrient signaling by reactive oxygen species in the cells. Maternal intake of eicosapentaenoic acid (EPA), present in fish oil, during fetal life improves signaling by inhibiting formation of AGEs.

3. n-3 unsaturated fatty acid intake issues for pregnant women

Our experiments with the GDM rat model revealed that EPA improves the intrauterine hyperglycemic environment during pregnancy. EPA is one of the fatty acids transported from the mother to the fetus via the placenta [41, 42] and is a very important essential fatty acid for the fetus [43–45]. However, EPA is known to have anticoagulant and thromboprophylaxis effects [46–49]. Therefore, very high intakes or levels of EPA may cause health problems, such as hemorrhage, prolonged gestation, or premature delivery, and thus intakes in pregnant women are being examined from various perspectives [50, 51]. In addition, given marine pollution in many areas, EPA and DHA extracted from fish oil have been found to be contaminated with environmental contaminants, such as methylmercury and polychlorinated biphenyls, which are known to cause neurological disorders when ingested by pregnant women and which can adversely affect the fetus [45, 50–56]. Therefore, it is recommended that EPA derived from marine products with high methylmercury concentrations, such as tuna and swordfish, be avoided.

4. Palmitoleic acid as n-7 unsaturated fatty acid

Our research using GDM model rats revealed that intrauterine hyperglycemia causes chronic inflammation in cardiomyocytes and various organs during fetal development due to the oxidative stress caused by excessive AGEs and that this causes disease development in the offspring. It was also found that maternal intake of EPA, an n-3 unsaturated fatty acid, during the fetal period ameliorates this chronic inflammation. However, given the concerns about the intake of EPA by pregnant women in terms of a bleeding tendency during delivery, we have been searching for alternative food functional ingredients to replace EPA. Palmitoleic acid, an n-7 unsaturated fatty acid, has recently been suggested to improve type 2 diabetes and to have anti-inflammatory effects. It has been called a lipokine because of its versatile physiological functions [57, 58]. Here, we describe the physiological properties of palmitoleic acid as a functional lipid that has been elucidated to date.

Among the unsaturated fatty acids, palmitoleic acid (C16:1, n-7; also known as 9-hexadecenoic acid) is a fatty acid found in blood and tissues, particularly in the adipose tissue and liver [59], as well as in human milk [60]. Palmitoleic acid is an n-7 monounsaturated fatty acid that is also naturally present in plants and fish oils [61, 62]. The classification of unsaturated fatty acids and the position of n-7 monounsaturated fatty acids are shown in **Figures 2** and **3**. In the human body, palmitoleic acid exists as cis and trans isomers. Cis-palmitoleic acid (cis-C16:1n-7, CPA) is synthesized from palmitic acid (C16:0), a saturated fatty acid produced in the body, in addition to normal dietary intake), which is synthesized mainly in the liver by stearoyl-CoA desaturase 1 (SCD1) (Δ9 desaturase) and is incorporated into triglycerides and other products. A portion is further converted to cis-vaccenic acid (cis-C18:1n-7) by an elongase-mediated elongation reaction [63]. Trans-palmitoleic acid (trans-C16:1n-7, TPA) is not biosynthesized from saturated fatty acids in the body and is mainly derived from trans fatty acids found in dairy and in ruminants [64, 65]. Ruminant trans-vaccenic acid (trans-C18:1n-7) is also biosynthesized endogenously in the human body by a chain-shortening reaction [66]. The metabolism of CPA and TPA as palmitoleic acid in the body is shown in **Figure 2**. Fish oil and macadamia nut oil

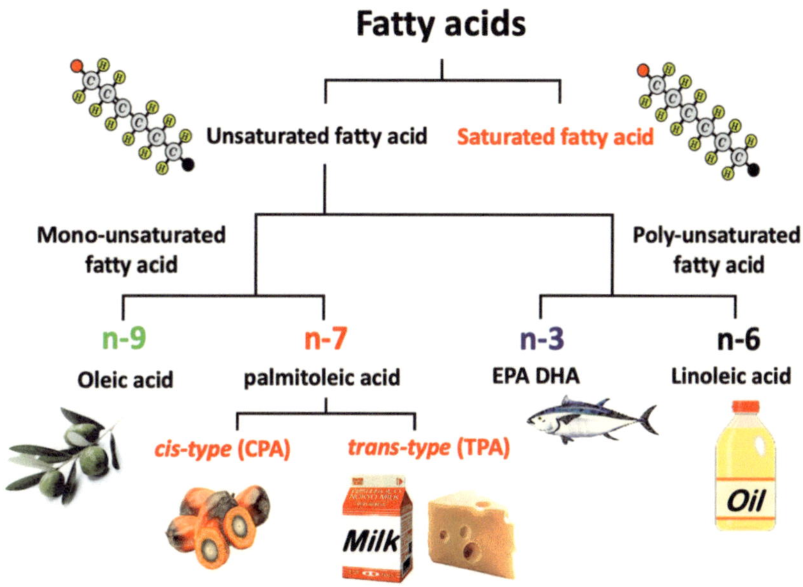

Figure 2.
Classification of fatty acids. Fatty acids are broadly classified into unsaturated and saturated fatty acids, depending on whether the molecule contains or does not contain double bonds between the carbons. Saturated fatty acids are abundant in meat and milk and include palmitic acid (16:0) and stearic acid (18:0). Unsaturated fatty acids are further classified into monounsaturated fatty acids, which have one double bond, and polyunsaturated fatty acids, which contain two or more double bonds. Among monounsaturated fats, oleic acid (18:1), found in olive oil, which has one double bond on the ninth carbon in the chain, is termed an n-9 unsaturated fatty acid. Palmitoleic acid (16:1), a fatty acid with one double bond on the seventh carbon, is referred to as an n-7 unsaturated fatty acid. Palmitoleic acid is further classified into two structural isomers: cis-type palmitoleic acid (CPA), which is abundant in meat fats and oils and in sea buckthorn fruit oil; and trans-type palmitoleic acid (TPA), which is found in milk and dairy products. Polyunsaturated fatty acids include eicosapentaenoic acid (20:5, EPA) and docosahexaenoic acid (20:6, DHA), which are found in fish oil, with the first double bond found on the third carbon from the methyl terminal of the carbon chain. Linoleic acid (18:2), a polyunsaturated fatty acid abundant in plant oil, has its first double bond on the sixth carbon from the methyl end of the carbon chain.

are dietary sources with higher amounts of CPA than other vegetable oils [67–69]. In addition, the oil of the fruit of sea buckthorn (*Hippophae rhamnoides L.*), which is widely grown in Asia, Europe, and Canada, contains about 40% palmitoleic acid [70, 71]. TPA is found in full-fat dairy products, and plasma concentrations of TPA correlate with the intake of full-fat dairy products and butter [72, 73]. Presumably, TPA is typically obtained by ingesting full-fat dairy products. Palmitoleic acid derived from food and its metabolism in the body are shown in **Figure 4**.

5. Palmitoleic acid as a functional lipid

In humans, palmitoleic acid is abundant in the muscle, liver, and adipose tissue, although the content of palmitoleic acid in the adipose tissue decreases with age [74, 75]. Cis-palmitoleic acid, an adipose-derived lipid hormone (lipokine), is intrinsically synthesized by SCD-1 in adipocytes through palmitic acid and acts in the liver and skeletal muscle to improve insulin sensitivity and metabolism in animal models [58, 76]. In addition, circulating levels of palmitoleic acid in humans are strongly correlated with insulin action [77]. Interestingly, the peroxisome proliferator-activated

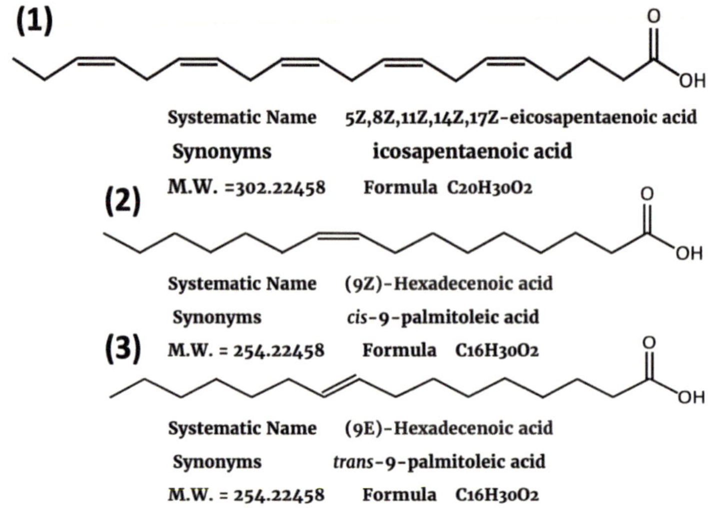

Figure 3.
Structural formulas for three functional lipids. (1) Eicosapentaenoic acid (EPA), an n-3 unsaturated fatty acid. (2) Cis-palmitoleic acid (CPA), an n-7 unsaturated fatty acid. (3) trans palmitoleic acid (TPA), an n-7 unsaturated fatty acid.

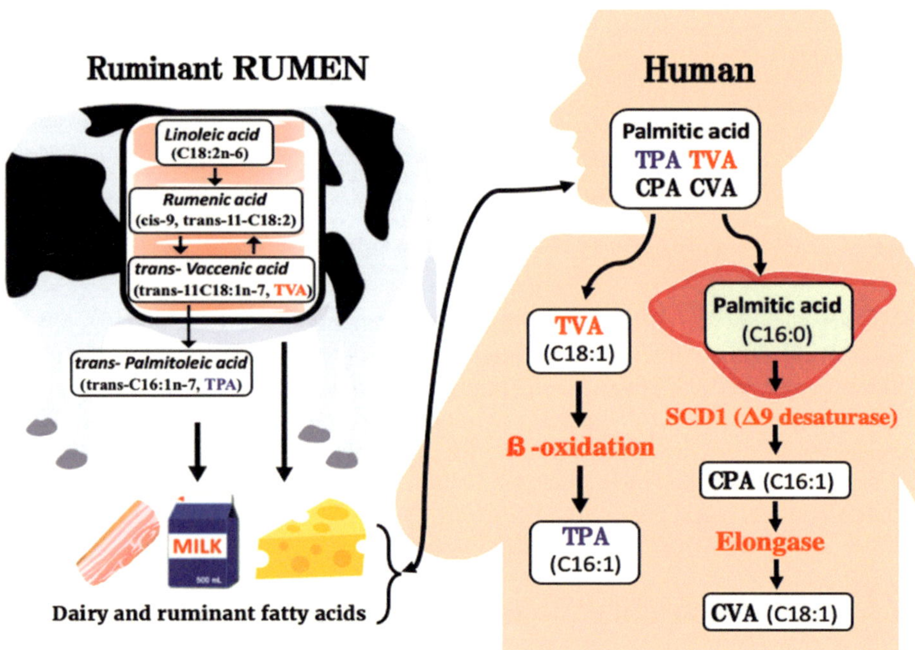

Figure 4.
Palmitoleic acid as a food and its metabolism in the human body. Palmitoleic acid (CPA) and trans-palmitoleic acid (TPA) have different routes of intake and metabolism in the body. In the human body, a double bond is introduced into palmitic acid, which is biosynthesized via de novo fatty acid synthesis in the liver and other organs, by SCD1 to form CPA. TPA cannot be biosynthesized in the human body and can only be obtained from milk and dairy products of ruminant origin. In ruminants, the gut bacteria biosynthesize conjugated linoleic acid, rumenic acid, from linoleic acid and store it as trans-vaccenic acid (TVA) in the body's adipose tissue, etc. TVA is further converted to TPA in the mammary gland and secreted into ruminant milk. Another route for obtaining TPA is through ingestion of TVA in the fat of pork and beef, which is partially converted to TPA by β-oxidation in the human body.

receptors (PPARs) are nuclear receptors that regulate DNA transcription, and three types are known: PPARα, PPARγ, and PPARδ. The oral administration of palmitoleic acid has been shown to suppress feeding by increasing cholecystokinin levels without involving this PPARα pathway [78]; moreover, it decreases phosphorylation of NF-κB p65 in mouse liver macrophages and inhibits the expression of inflammatory cytokines [79]. Palmitoleic acid also suppresses gene expression of palmitin-induced inflammatory cytokines via AMP-activated protein kinase, indicating a mechanism by which it exerts an anti-inflammatory effect [80]. Topical administration of palmitoleic acid has been reported to heal wounds due to its anti-inflammatory effects [81]. In humans, an association between dietary-derived and blood palmitoleic acid levels and a decreased incidence of diabetes, risk of cardiovascular disease, and inflammatory status have been suggested [77, 82, 83]. Thus, palmitoleic acid is thought to have an inhibitory effect on the development of diabetes by improving insulin resistance and increasing pancreatic beta cell proliferation and insulin secretion in animal and cellular models [84–89].

Thus, in summary, many studies have shown that palmitoleic acid, as a functional lipid, is effective in preventing the onset of lifestyle-related diseases and has anti-inflammatory, anti-diabetic, and anti-obesity effects. Palmitoleic acid has two structural isomers: the cis-type CPA and the trans-type TPA. The former is biosynthesized from palmitic acid, a saturated fatty acid that causes inflammation and organ stress in skeletal and cardiac muscle, by introducing double bonds and chain-elongation reactions, while the latter cannot be synthesized in the human body and is derived from dietary fatty acids found in meat and milk of ruminant origin (**Figure 4**).

Little research has been done on the differential effects of the CPA and TPA isomers on biological functions. Recently, we found that high-glucose-exposed myoblasts show increased phosphorylation levels of MAPK/ERK1/2, ROS production, and inflammatory signals; however, their treatment with TPA and EPA suppressed both ROS production and inflammation. In contrast, the opposite was found with a CPA treatment, which increased ROS production and demonstrated higher cytotoxicity than TPA [90]. Thus, CPA and TPA are functional lipids that may affect the organism differently.

Recently, an increasing number of clinical studies have focused on TPA derived from dairy products. In the Cardiovascular Health Study, a large prospective cohort study in the United States, it was found that older adults with a higher percentage of TPA in their blood total fat content had lower insulin resistance and a significantly lower incidence of type 2 diabetes [91]. The Multi-Ethnic Study of Atherosclerosis (MESA) found that TPA in blood was associated with elevated LDL cholesterol, but reduced triglycerides, fasting insulin, blood pressure, and diabetes incidence [92]. Two cohort studies, the Nurses' Health Study and the Health Professionals Follow-Up Study, of 3333 adults without diabetes aged 30–75 years, found that higher plasma milk-derived TPA levels were associated with a lower risk of developing diabetes mellitus [93]. In a meta-analysis examining the association of saturated and trans-unsaturated fat intake with all-cause mortality, cardiovascular disease, coronary heart disease, ischemic stroke, and type 2 diabetes, TPA intake from a ruminant-derived diet was inversely associated with type 2 diabetes [94]. Another study showed that, as a potential biomarker of milk fat intake, high levels of TPA in circulating blood or adipose tissue were associated with a lower risk of developing type 2 diabetes [95].

In a nonclinical basic research study using db/db mice, TPA was shown to ameliorate hypercholesterolemia by reducing serum cholesterol, low-density lipoprotein, high-density lipoprotein, and hepatic free cholesterol levels, while CPA had none of these effects [96]. TPA also inhibited cholesterol absorption from the intestinal tract by markedly reducing the expression of intestinal Niemann-Pick C1-like 1 protein. However, histological examination of the liver showed that CPA more effectively ameliorated hepatic lipidosis than did TPA. These results suggest that TPA and CPA may prevent hypercholesterolemia by different mechanisms [96]. In a mouse model of diet-induced obesity, TPA prevented weight gain caused by a high-fat diet, reduced visceral adipose tissue weight and adipocyte size, and increased expression of lipolysis-related genes [97]. In vascular endothelial cells and hepatocytes, TPA reduced TNF-α-induced inflammation [98], and in experiments with β-cell lines, TPA and the branched-chain fatty acid 15-methylhexadecanoic acid (iso 17:0) increased PPARγ activity by about two fold and enhanced β-cell function, thereby improving insulin resistance. TPA has been reported to improve insulin resistance by increasing β-cell function [99]. Palmitic acid and CPA decreased hepatocyte viability at concentrations above 1 mM, whereas TPA treatment had a cell proliferative effect: TPA had a beneficial effect on hepatocyte survival signals by activating sirtuin 1 and inducing PPARα activity [100].

In the Ludwigshafen Risk and Cardiovascular Health (LURIC) clinical study of the association between trans fatty acids and mortality, TPA was associated with a reduced risk of death due to cardiovascular causes, but this effect was not observed with industrially produced hydrogenated fats and oils [101]. The effects of palmitoleic acid were examined using isoproterenol-induced myocardial damage model mice and primary mouse cardiomyocytes, and the results showed that palmitoleic acid may have a protective effect against cardiac fibrosis and inflammation by regulating PPAR-specific signaling pathways in the heart [102].

6. Conclusions

When a woman has diabetes during pregnancy, her hyperglycemia creates a hyperglycemic intrauterine environment for the fetus via the placenta. Exposure to this environment during the fetal period is closely related to the future development of cardiac and neurological diseases in the offspring. We have been conducting research to find food ingredients that reduce insulin resistance in pregnant women. While EPA has anti-diabetic effects, there are concerns about the active use of EPA in pregnant women due to its anticoagulant effect and due to the marine pollution of fish oil, which is a primary source of EPA supplementation. In this paper, we propose the use of palmitoleic acid, an n-7 unsaturated fatty acid, which has anti-obesity, lipid metabolism improving, and cardioprotective effects similar to those of EPA. Palmitoleic acid is classified into cis-type CPA and trans-type TPA. Further basic molecular studies are needed to clarify the differences in the mechanisms of action of these structural isomers. In particular, TPA has beneficial effects on the cardiovascular system. However, the intake of TPA, which is only marginally present in dairy products, must be carefully considered from a mother-child nutritional perspective, as excessive intake of dairy products by the mother may lead to over-nutrition and allergic reactions in the child. Recently, a synthesis that converts CPA, which is abundant in macadamia nut oil and sea buckthorn fruit oil, to TPA has been developed, and research on palmitoleic acid-producing bacteria is underway. Thus, it is expected that TPA will be available as an inexpensive, safe, and accessible supplement in the future.

Acknowledgements

We gratefully acknowledge the work of past and present members of our laboratory. This work was supported in part by the JSPS KAKENHI Grants (nos. 22 K 11882, 20 K11611, and 18 K11136 to AN and RK), the Dairy Products Health Science Council and Japan Dairy Association (to AN), and the Research Program of Jissen Women's University (to AN).

Conflict of interest

The authors declare no conflict of interest.

Author details

Akio Nakamura[1*], Hikari Nakamura[2] and Ritsuko Kawaharada[3]

1 Faculty of Human Life Sciences, Department of Molecular Nutrition, Jissen Women's University, Hino, Tokyo, Japan

2 Laboratory of Membrane Trafficking Mechanisms, Department of Integrative Life Sciences, Graduate School of Life Sciences, Tohoku University, Sendai, Japan

3 Department of Health and Nutrition, Takasaki University of Health and Welfare, Takasaki, Gunma, Japan

*Address all correspondence to: nakamura-akio@jissen.ac.jp

References

[1] Barker DJ, Osmond C. Infant mortality, childhood nutrition, and ischaemic heart disease in England and Wales. Lancet. 1986;**1**:1077-1081. DOI: 10.1016/s0140-6736(86)91340-1

[2] Barker DJ, Winter PD, Osmond C. Weight in infancy and death from ischaemic heart disease. Lancet. 1989;**2**:577-580

[3] Barker DJ. The fetal and infant origins of adult disease. British Medical Journal. 1990;**30**:1111. DOI: 10.1136/bmj.301.6761.1111

[4] Barker DJ, Gluckman PD, Godfrey KM. Fetal nutrition and cardiovascular disease in adult life. Lancet. 1993;**341**:938-941. DOI: 10.1016/0140-6736(93)91224-a

[5] Barker DJ. The fetal and infant origins of disease. European Journal of Clinical Investigation. 1995;**25**(7):457-463. DOI: 10.1111/j.1365-2362.1995.tb01730.x

[6] Gluckman PD, Hanson MA. Living with the past:Evolution, development, and patterns of disease. Science. 2004;**305**:1733-1736. DOI: 10.1126/science.1095292

[7] Gluckman PD, Hanson MA. Developmental origins of disease paradigm:A mechanistic and evolutionary perspective. Pediatric Research. 2004;**56**:311-317. DOI: 10.1203/01.PDR.0000135998.08025.FB

[8] Hales CN, Barker DJ. Type 2 (non-insulin-dependent) diabetes mellitus: The thrifty phenotype hypothesis. International Journal of Epidemiology. 2012;**42**:1215-1222

[9] Arima Y, Fukuoka H. Developmental origins of health and disease theory in cardiology. Journal of Cardiology. 2020;**76**:14-17. DOI: 10.1016/j.jjcc.2020.02.003. Epub 2020 Feb 27

[10] Hoffman DJ, Powell TL, Barrett ES. Developmental origins of metabolic diseases. Physiological Reviews. 2021;**101**(3):739-795. DOI: 10.1152/physrev.00002.2020

[11] McAllister EJ, Dhurandhar NV, Keith SW. Ten putative contributors to the obesity epidemic. Critical Reviews in Food Science and Nutrition. 2009;**49**(10):868-913. DOI: 10.1080/10408390903372599

[12] Popkin BM. Global nutrition dynamics: The world is shifting rapidly toward a diet linked with noncommunicable diseases. The American Journal of Clinical Nutrition. 2006;**84**(2):289-298. DOI: 10.1093/ajcn/84.1.289

[13] Hunt KJ, Schuller KL. The increasing prevalence of diabetes in pregnancy. Obstetrics and Gynecology Clinics of North America. 2007;**34**(2):173-199, vii. DOI: 10.1016/j.ogc.2007.03.002

[14] Barnes-Powell LL. Infants of diabetic mothers: The effects of hyperglycemia on the fetus and neonate. Neonatal Network. 2007;**26**:283-290. DOI: 10.1891/0730-0832.26.5.283

[15] Kong L, Nilsson IAK, Gissler M. Associations of maternal diabetes and body mass index with offspring birth weight and prematurity. JAMA Pediatrics. 2019;**173**:371-378. DOI: 10.1001/jamapediatrics.2018.5541

[16] Clausen TD, Mathiesen ER, Hansen T. High prevalence of type 2 diabetes and pre-diabetes in adult

offspring of women with gestational diabetes mellitus or type 1 diabetes: The role of intrauterine hyperglycemia. Diabetes Care. 2008;**31**:340-346

[17] Ornoy A, Becker M, Weinstein-Fudim L. Diabetes during pregnancy: A maternal disease complicating the course of pregnancy with long-term deleterious effects on the offspring. A clinical review. International Journal of Molecular Sciences. 2021;**22**(6):2965. DOI: 10.3390/ijms22062965

[18] Choudhury AA, Devi RV. Gestational diabetes mellitus—A metabolic and reproductive disorder. Biomedicine and Pharmacotherapy. 2021;**143**:112183. DOI: 10.1016/j.biopha.2021.112183. Epub 2021 Sep 21

[19] Nasu R, Seki K, Nara M. Effect of a high-fat diet on diabetic mother rats and their offspring through three generations. Endocrine Journal. 2007;**54**:563-569. DOI: 10.1507/endocrj.k06-175

[20] Perdomo L, Beneit N, Otero YF. Protective role of oleic acid against cardiovascular insulin resistance and in the early and late cellular atherosclerotic process. Cardiovascular Diabetology. 2015;**14**:75. DOI: 10.1186/s12933-015-0237-9

[21] Kwon B, Lee HK, Querfurth HW. Oleate prevents palmitate-induced mitochondrial dysfunction, insulin resistance and inflammatory signaling inneuronal cells. Biochimica et Biophysica Acta. 2014;**1843**:1402-1413. DOI: 10.1016/j.bbamcr.2014.04.004

[22] Salvado L, Coll T, Gomez-Foix AM. Oleate prevents saturated-fatty-acid-induced ER stress, inflammation and insulin resistance in skeletal muscle cells through an AMPK-dependent mechanism. Diabetologia. 2013;**56**:1372-1382. DOI: 10.1007/s00125-013-2867-3

[23] Gao D, Griffiths HR, Bailey CJ. Oleate protects against palmitate-induced insulin resistance in L6 myotubes. British Journal of Nutrition. 2009;**102**:1557-1563. DOI: 10.1017/S0007114509990948

[24] Yang L, Guan G, Lei L. Oxidative and endoplasmic reticulum stresses are involved in palmitic acid-induced H9c2 cell apoptosis. Bioscience Reports. 2019;**39**(5):BSR20190225. DOI: 10.1042/BSR20190225

[25] Gui T, Li Y, Zhang S. Docosahexaenoic acid protects against palmitate-induced mitochondrial dysfunction in diabetic cardiomyopathy. Biomedicine and Pharmacotherapy;**128**:110306. DOI: 10.1016/j.biopha.2020.110306

[26] Park S, Park Y. Effects of dietary fish-oil and trans fat on rat aorta histopathology and cardiovascular risk markers. Nutrition Research and Practice. 2009;**3**:102-107. DOI: 10.4162/nrp.2009.3.2.102

[27] Kromhout D, Geleijnse JM, Goede J. n-3 fatty acids, ventricular arrhythmia-related events, and fatal myocardial infarction in postmyocardial infarction patients with diabetes. Diabetes Care. 2011;**34**:2515-2520. DOI: 10.2337/dc11-0896

[28] Nasu-Kawaharada R, Nakamura A, Kakarla S. A maternal diet rich in fish oil may improve cardiac Akt-related signaling in the offspring of diabetic mother rats. Nutrition. 2013;**29**:688-692. DOI: 10.1016/j.nut.2012.11.017

[29] Watanabe T, Ando K, Daidoji H. CHERRY study investigators. A randomized controlled trial of eicosapentaenoic acid in patientswith

coronary heart disease on statins. Journal of Cardiology. 2017;**70**:537-544. DOI: 10.1016/j.jjcc.2017.07.007

[30] Borow KM, Nelson JR, Mason RP. Biologic plausibility, cellular effects, and molecular mechanisms of eicosapentaenoic acid (EPA) in atherosclerosis. Atherosclerosis. 2015;**242**:357-366. DOI: 10.1016/j.atherosclerosis.2015.07.035

[31] Cawood AL, Ding R, Napper FL. Eicosapentaenoic acid (EPA) from highly concentrated n-3 fatty acid ethyl esters is incorporated into advanced atherosclerotic plaques and higher plaque EPA is associated with decreased plaque inflammation and increased stability. Atherosclerosis. 2010;**212**:252-259. DOI: 10.1016/j.atherosclerosis.2010.05.022

[32] Nishio R, Shinke T, Otake H. Stabilizing effect of combined eicosapentaenoic acid and statin therapy on coronary thin-cap fibroatheroma. Atherosclerosis. 2014;**234**:114-119. DOI: 10.1016/j.atherosclerosis.2014.02.025

[33] Nosaka K, Miyoshi T, Iwamoto M. Early initiation of eicosapentaenoic acid and statin treatment is associated with better clinical outcomes than statin alone in patients with acute coronary syndromes: 1-year outcomes of a randomized controlled study. International Journal of Cardiology. 2017;**228**:173-179. DOI: 10.1016/j.ijcard.2016.11.105

[34] Kawaharada R, Masuda H, Chen Z. Intrauterine hyperglycemia-induced inflammatory signalling via the receptor for advanced glycation end products in the cardiac muscle of the infants of diabetic mother rats. European Journal of Nutrition. 2018;**57**:2701-2712. DOI: 10.1007/s00394-017-1536-6

[35] Qiu HY, Hou NN, Shi JF. Comprehensive overview of human serum albumin glycation in diabetes mellitus. World Journal of Diabetes. 2021;**12**:1057-1069. DOI: 10.4239/wjd.v12.i7.1057

[36] Hegab Z, Gibbons S, Neyses L. Role of advanced glycation end products in cardiovascular disease. World Journal of Cardiology. 2012;**4**:90-102. DOI: 10.4330/wjc.v4.i4.90

[37] Akhter F, Chen D, Akhter A. High dietary advanced glycation end products impair mitochondrial and cognitive function. Journal of Alzheimer's Disease. 2020;**76**:165-178. DOI: 10.3233/JAD-191236

[38] Nishad R, Tahaseen V, Kavvuri R. Advanced-glycation end-products induce podocyte injury and contribute to proteinuria. Frontiers in Medicine. 2021;**8**:685447. DOI: 10.3389/fmed.2021.685447

[39] Rungratanawanich W, Qu Y, Wang X. Advanced glycation end products (AGEs) and other adducts in aging-related diseases and alcohol-mediated tissue injury. Experimental & Molecular Medicine. 2021;**53**:168-188. DOI: 10.1080/2162402X.2021.1874159

[40] Petriv N, Neubert L, Vatashchuk M. Increase of -Dicarbonyls in liver and receptor for advanced glycation end products on immune cells are linked to nonalcoholic fatty liver disease and liver Cancer. Oncoimmunology. 2021;**10**:1874159. DOI: 10.1080/2162402X.2021.1874159

[41] Lewis RM, Childs CE, Calder PC. New perspectives on placental fatty acid transfer. Prostaglandins, Leukotrienes and Essential Fatty Acids. 2018;**138**:24-29. DOI: 10.1016/j.plefa.2018.10.001

[42] Devarshi PP, Grant RW, Ikonte CJ. Maternal Omega-3 nutrition, placental transfer and fetal brain development in gestational diabetes and preeclampsia. Nutrients. 2019;**11**(5):1107. DOI: 10.3390/nu11051107

[43] Shrestha N, Sleep SL, Cuffe JSM. Role of omega-6 and omega-3 fatty acids in fetal programming. Clinical and Experimental Pharmacology and Physiology. 2020;**47**(5):907-915. DOI: 10.1111/1440-1681.13244

[44] Greenberg JA, Bell SJ, Ausdal WV. Omega-3 fatty acid supplementation during pregnancy. Reviews in Obstetrics and Gynecology. 2008;**1**(4):162-169

[45] Coletta JM, Bell SJ, Roman AS. Omega-3 fatty acids and pregnancy. Reviews in Obstetrics and Gynecology. 2010;**3**(4):163-171

[46] Golanski J, Szymanska P, Rozalski M. Effects of Omega-3 polyunsaturated fatty acids and their metabolites on Haemostasis-current perspectives in cardiovascular disease. International Journal of Molecular Sciences. 2021;**22**(5):2394. DOI: 10.3390/ijms22052394

[47] Larson MK, Tormoen GW, Weaver LJ. Exogenous modification of platelet membranes with the omega-3 fatty acids EPA and DHA reduces platelet procoagulant activity and thrombus formation. American Journal of Physiology. 2013;**304**(3):C273-C279. DOI: 10.1152/ajpcell.00174.2012

[48] McEwen B, Morel-Kopp MC, Tofler G. The effect of Omega-3 polyunsaturated fatty acids on fibrin and thrombin generation in healthy subjects and subjects with cardiovascular disease. Seminars in Thrombosis and Hemostasis. 2015;**41**:315-322. DOI: 10.1055/s-0034-1395352

[49] Yokoyama M, Origasa H, Matsuzaki M. Effects of eicosapentaenoic acid on major coronary events in hypercholesterolaemic patients (JELIS): A randomised open-label, blinded endpoint analysis.; Japan EPA lipid intervention study (JELIS) investigators. Lancet. 2007;**369**(9567):1090-1098. DOI: 10.1016/S0140-6736(07)60527-3

[50] Schacky CV. Omega-3 fatty acids in pregnancy-the case for a target Omega-3 index. Nutrients. 2020;**12**(4):898. DOI: 10.3390/nu12040898

[51] Simmonds LA, Sullivan TR, Skubisz M. Omega-3 fatty acid supplementation in pregnancy-baseline omega-3 status and early preterm birth: Exploratory analysis of a randomised controlled trial. BJOG : An International Journal of Obstetrics and Gynaecology. 2020;**127**(8):975-981. DOI: 10.1111/1471-0528.16168

[52] Keating MH, Mahaffey KR, Schoemy R. Mercury study report to congress. In: Executive Summary. EPA-452/R-97-003. Vol. I. Washington DC: Environmental Protection Agency; 1997

[53] Yamaguchi N, Gazzard D, Scholey G. Concentrations and hazard assessment of PCBs, organochlorine pesticides and mercury in fish species from the upper Thames: River pollution and its potential effects on top predators. Chemosphere. 2003;**50**:265-273

[54] Guallar E, Sanz-Gallardo MI, Veer P. Mercury, fish oils, and the risk of myocardial infarction. The New England Journal of Medicine. 2002;**347**:1747-1754

[55] Virtanen JK, Voutilainen S, Rissanen TH. Mercury, fish oils, and risk of acute coronary events and cardiovascular disease, coronary heart disease, and all-cause mortality in men in eastern Finland. Arteriosclerosis,

Thrombosis, and Vascular Biology. 2005;**25**:228-233. DOI: 10.1161/01.ATV.0000150040.20950.61

[56] Sacks D, Baxter B, Campbell BCV. Multisociety consensus quality improvement revised consensus statement for endovascular therapy of acute ischemic stroke. International Journal of Stroke. 2018;**13**(6):612-632. DOI: 10.1177/1747493018778713

[57] Yang ZH, Miyahara H, Hatanaka A. Chronic administration of palmitoleic acid reduces insulin resistance and hepatic lipid accumulation in KK-Ay Mice with genetic type 2 diabetes. Lipids in Health and Disease. 2011;**10**:120. DOI: 10.1186/1476-511X-10-120

[58] Cao H, Gerhold K, Mayers JR. Identification of a lipokine, a lipid hormone linking adipose tissue to systemic metabolism. Cell. 2008;**134**:933-944. DOI: 10.1016/j.cell.2008.07.048

[59] Hodson L, Skeaff CM, Fielding BA. Fatty acid composition of adipose tissue and blood in humans and its use as a biomarker of dietary intake. Progress in Lipid Research. 2008;**47**(5):348-380. DOI: 10.1016/j.plipres.2008.03.003

[60] Ogunleye A, Fakoya AT, Niizeki S. Fatty acid composition of breast milk from Nigerian and Japanese women. Journal of Nutritional Science and Vitaminology. 1991;**37**(4):435-442. DOI: 10.3177/jnsv.37.435

[61] Yang B, Kallio HP. Fatty acid composition of lipids in sea buckthorn (Hippophaë rhamnoides L.) berries of different origins. Journal of Agricultural and Food Chemistry. 2001;**49**(4):1939-1947. DOI: 10.1021/jf001059s

[62] Fard AM, Turner AG, Willett GD. High-resolution electrospray-ionization fourier-transform ion cyclotron resonance and gas chromatography-mass spectrometry of macadamia nut oil. Australian Journal of Chemistry. 2003;**56**(5):499-508. DOI: 10.1071/CH03038

[63] Gong J, Campos H, McGarvey S. Adipose tissue palmitoleic acid and obesity in humans: Does it behaveas a lipokine? The American Journal of Clinical Nutrition. 2011;**93**:186-191

[64] Guillocheau E, Garcia C, Drouin G. Retroconversion of dietary trans-vaccenic (trans-C18:1 n-7) acid to trans-palmitoleic acid (trans-C16:1 n-7): Proof of concept and quantification in both cultured rat hepatocytes and pregnant rats. The Journal of Nutritional Biochemistry. 2019;**63**:19-26. DOI: 10.1016/j.jnutbio.2018.09.010

[65] Destaillats F, Wolff RL, Precht D. Study of individual trans- and cis-16:1 isomers in cow, goat, and ewe cheese fats by gas-liquid chromatography with emphasis on the trans-delta3 isomer. Journal of Lipids. 2000;**35**:1027-1032. DOI: 10.1007/s11745-000-0614-y

[66] Jaudszus A, Kramer R, Pfeuffer M. Trans-Palmitoleic acid arises endogenously from dietary vaccenic acid. The American Journal of Clinical Nutrition. 2014;**99**:431-435. DOI: 10.3945/ajcn.113.076117

[67] Garg ML, Blake RJ, Wills RB. Macadamia nut consumption lowers plasma total and LDL cholesterol levels in hypercholesterolemic men. Journal of Nutrition. 2003;**133**(4):1060-1063. DOI: 10.1093/jn/133.4.1060

[68] Maguire LS, O'Sullivan SM, Galvin K. Fatty acid profile, tocopherol, squalene and phytosterol content of walnuts, almonds, peanuts, hazelnuts and the macadamia nut. International Journal of Food Sciences

and Nutrition. 2004;**55**(3):171-178. DOI: 10.1080/09637480410001725175

[69] Borompichaichartkul C, Luengsode K, Chinprahast N. Improving quality of Macadamia nut (Macadamia integrifolia) through the use of hybrid drying. Journal of Food Process Engineering. 2009;**93**:348-353. DOI: 10.1016/j.jfoodeng.2009.01.035

[70] Fatima T, Snyder CL, Schroeder WR. Fatty acid composition of developing sea buckthorn (Hippophae rhamnoides L.) berry and the transcriptome of the mature seed. PLoS One. 2012;7(4):e34099. DOI: 10.1371/journal.pone.0034099

[71] Ciesarová Z, Murkovic M, Cejpek K. Why is sea buckthorn (Hippophae rhamnoides L.) so exceptional? A review. Food Research International. 2020;**133**:109170. DOI: 10.1016/j.foodres.2020.109170

[72] Micha R, King IB, Lemaitre RN. Food sources of individual plasma phospholipid trans fatty acid isomers: The cardiovascular health study. The American Journal of Clinical Nutrition. 2010;**91**:883-893. DOI: 10.3945/ajcn.2009.28877

[73] Pranger IG, Corpeleijn E, Muskiet FAJ. Circulating fatty acids as biomarkers of dairy fat intake: Data from the lifelines biobank and cohort study. Biomarkers. 2019;**24**:1-39. DOI: 10.1080/1354750X.2019.1583770

[74] McLaren DS, Ajans ZA, Awdeh Z. Composition of human adipose tissue, with special reference to site and age differences. American Journal of Clinical Nutrition. 1965;**17**:171-176. DOI: 10.1093/ajcn/17.3.171

[75] Pinnick KE, Neville MJ, Fielding BA. Gluteofemoral adipose tissue plays a major role in production of the lipokine palmitoleate in humans. Diabetes. 2012;**61**:1399-1403. DOI: 10.2337/db11-1810

[76] Souza CO, Teixeira AA, Lima EA. Palmitoleic acid (n-7) attenuates the immunometabolic disturbances caused by a high-fat diet independently of PPARα. Mediators of Inflammation. 2014;**2014**:582197. DOI: 10.1155/2014/582197

[77] Stefan N, Kantartzis K, Celebi N. Circulating palmitoleate strongly and independently predicts insulin sensitivity in humans. Diabetes Care. 2010;**33**:405-407. DOI: 10.2337/dc09-0544

[78] Yang ZH, Takeo J, Katayama M. Oral administration of omega-7 palmitoleic acid induces satiety and the release of appetite-related hormones in male rats. Appetite. 2013;**65**:1-7. DOI: 10.1016/j.appet.2013.01.009

[79] Guo X, Li H, Xu H. Palmitoleate induces hepatic steatosis but suppresses liver inflammatory response in mice. PLoS One. 2012;**7**:e39286. DOI: 10.1371/journal.pone.0039286

[80] Chan KL, Pillon NJ, Sivaloganathan DM. Journal of Biological Chemistry. 2015;**290**(27):16979-16988. DOI: 10.1074/jbc.M115.646992. Epub 2015 May 18

[81] Weimann E, Silva MBB, Murata GM, Bortolon JR, Dermargos A, Curi R, et al. Topical anti-inflammatory activity of Palmitoleic acid ImprovesWound healing. PLoS One. 2018;**13**:e0205338

[82] Bernstein AM, Roizen MF, Martinez L. Purified palmitoleic acid for the reduction of high-sensitivity C-reactive protein and serum lipids: A double-blinded, randomized, placebo

controlled study. Journal of Clinical Lipidology. 2014;**8**:612-617

[83] Morse N. Lipid-lowering and anti-inflammatory effects of Palmitoleic acid: Evidence from preclinical and epidemiological studies. Lipid Technology. 2015;**27**:107-111. DOI: 10.1002/lite.201500019

[84] Maedler K, Oberholzer J, Bucher P. Monounsaturated fatty acids prevent the deleterious effects of palmitate and high glucose on human pancreatic beta-cell turnover and function. Diabetes. 2003;**52**:726-733. DOI: 10.2337/diabetes.52.3.726

[85] Dimopoulos N, Watson M, Sakamoto K. Differential effects of palmitate and palmitoleate on insulin action and glucose utilization in rat L6 skeletal muscle cells. Biochemical Journal. 2006;**399**:473-481

[86] Welters HJ, Diakogiannaki E, Mordue JM. Differential protective effects of palmitoleic acid and cAMP on caspase activation and cell viability in pancreatic beta-cells exposed to palmitate. Apoptosis. 2006;**11**:1231-1238. DOI: 10.1007/s10495-006-7450-7

[87] Tsuchiya Y, Hatakeyama H, Emoto N. Palmitate-induced down-regulation of sortilin and impaired GLUT4 trafficking in C2C12 myotubes. Journal of Biological Chemistry. 2010;**285**:34371-34381

[88] Matthan NR, Dillard A, Lecker JL. Effects of dietary palmitoleic acid on plasma lipoprotein profile and aortic cholesterol accumulation are similar to those of other unsaturated fatty acids in the F1B golden Syrian hamster. Journal of Nutrition. 2009;**139**:215-221

[89] Morgan NG, Dhayal S. Unsaturated fatty acids as cytoprotective agents in the pancreatic β-cell. Prostaglandins Leukot. Essential Fatty Acid. 2010;**82**:231-236. DOI: 10.1016/j.plefa.2010.02.018

[90] Tokunaga Y, Yoshizaki H, Toriumi A. Effects of omega-7 palmitoleic acids on skeletal muscle differentiation in a hyperglycemic condition. The Journal of Veterinary Medical Science. 2021;**83**:1369-1377. DOI: 10.1292/jvms.21-0309. Epub 2021 Jul 12

[91] Mozaffarian D, Cao H, King IB. Trans-palmitoleic acid, metabolic risk factors, and new-onset diabetes in U.S. adults: A cohort study. Annals of Internal Medicine. 2010;**153**:790-799. DOI: 10.7326/0003-4819-153-12-201012210-00005

[92] Mozaffarian D, Oliveira OMC, Lemaitre RN. Trans-Palmitoleic acid, other dairy fat biomarkers, and incident diabetes: The multi-ethnic study of atherosclerosis (MESA). The American Journal of Clinical Nutrition. 2013;**97**:854-861

[93] Yakoob MY, Shi P, Willett WC. Circulating biomarkers of dairy fat and risk of incident diabetes mellitus among US men and women in two large prospective cohorts. Circulation. 2016;**133**:1645-1654. DOI: 10.1161/CIRCULATIONAHA.115.018410

[94] Souza RJ, Mente A, Maroleanu A. Anand intake of saturated and trans unsaturated fatty acids and risk of all cause mortality, cardiovascular disease, and type 2 diabetes: Systematic review and meta-analysis of observational studies. British Medical Journal. 2015;**351**:1-16. DOI: 10.1136/bmj.h3978

[95] Imamura F, Fretts AM, Marklund M. Fatty acid biomarkers of dairy fat consumption and incidence of type 2 diabetes: A pooled analysis of prospective cohort studies. PLoS Medicine. 2018;**15**:1-18. DOI: 10.1371/journal.pmed.1002670

[96] Huang WW, Hong BH, Bai KK. Cis- and trans-Palmitoleic acid isomers regulate cholesterol metabolism in different ways. Frontiers in Pharmacology. 2020;**11**:602115. DOI: 10.3389/fphar.2020.602115

[97] Chávaro-Ortiz LI, Tapia BD, Rico-Hidalgo M. Trans-palmitoleic acid reduces adiposity via increased lipolysis in a rodent model of diet-induced obesity. British Journal of Nutrition. 2022;**127**:801-809. DOI: 10.1017/S0007114521001501

[98] Da Silva MS, Julien P, Bilodeau J-F. Trans fatty acids suppress TNFα-induced inflammatory gene expression in endothelial (HUVEC) and hepatocellular carcinoma (HepG2) cells. Lipids. 2017;**52**:315-325. DOI: 10.1007/s11745-017-4243-4

[99] Kraft J, Jetton T, Satish B. Dairy-derived bioactive fatty acids improve pancreatic ß-cell function. FASEB Journal. 2015;**29**:608-625

[100] Farokh Nezhad R, Nourbakhsh M, Razzaghy-Azar M. The effect of trans-palmitoleic acid on cell viability and sirtuin 1 gene expression in hepatocytes and the activity of peroxisome-proliferator-activated receptor-alpha. Journal of Research in Medical Sciences. 2020;**25**:105. DOI: 10.4103/jrms.JRMS_16_20

[101] Kleber ME, Delgado GE, Lorkowski S. Trans-fatty acids and mortality in patients referred for coronary angiography: The Ludwigshafen risk and cardiovascular health study. European Heart Journal. 2016;**37**:1072-1078

[102] Betz IR, Qaiyumi SJ, Goeritzer M. Cardioprotective effects of Palmitoleic acid (C16:1n7) in a mouse model of catecholamine-induced cardiac damage are mediated by PPAR activation. International Journal of Molecular Sciences. 2021;**22**:12695. DOI: 10.3390/ijms222312695.doi: 10.1093/eurheartj/ehv446

Section 5

Lipid Metabolic Disorders and Lipidomics

Chapter 9

Lipid Metabolic Defects and Lipid-Dependent Gating of Voltage-Gated Ion Channels

Qiu-Xing Jiang and Felix Chin

Abstract

Eukaryotic cells contain phospholipids and nonphospholipids. The latter lack phosphodiester groups in their head group regions. Lipid-dependent gating of voltage-gated ion channels represents a steady-state energetic effect of nonphospholipids in favoring the resting state of voltage-sensor domains (VSDs) of the channels. It suggests adaptation of ion channels to lipid compositions in their native niche and significant roles of low-to-intermediate affinity lipid-binding sites at the channels. The nonphospholipids include glycoglycerolipids, glycosphingolipids, ceramides, cholesterol or cholesterol esters, diacylglycerol (DAG), fatty acids, cation lipids, etc. Change in relative ratios of phospholipids to nonphospholipids can shift the energetic levels of the VSDs and the gating of these channels, which in turn may alter excitability in certain cells. It is expected that reduced relative abundance of nonphospholipids / phospholipids in plasma membranes may change resting transmembrane potential or gating transitions of voltage-gated Na or K channels. The net results will be a change in action potential firing at least in certain areas of an excitable cell. Such changes in the central nervous system (CNS) are anticipated to affect brain functions and contribute to early-onset neurological phenotypes observed in patients carrying lipid metabolic defects. We will describe the basics of lipid-dependent gating and review its projected links to phenotypes of monogenic lipid metabolic defects and related changes of lipid composition in cell membranes as well as altered neuronal excitability in CNS. However, lack of high-resolution techniques to measure lipid composition around individual channels in cell membranes has been limiting the studies of direct connections between lipid redistribution caused by metabolic defects and altered ion channel activities. Potential solutions will be described for future studies.

Keywords: lipid metabolic defects, lipid-dependent gating, nonphospholipids, voltage-gated ion channels, resting state, voltage sensor domains, activation, inactivation, low affinity lipid-protein interactions, steady-state energetics

1. Introduction

All living cells are thermodynamically off-equilibrium. As open systems, their identities and existence are demarcated by their plasma membranes, and they exchange materials and energy through these membranes [1]. Lipid metabolism basically covers exchange of lipid molecules or their constituents between a cell and its environment or between an organism and its surroundings as well as lipid distribution (or redistribution) and composition within a cell or an organism [2]. At the level of individual cells, lipids can be distributed among the plasma membranes, the intracellular membranes, and the nonlamellar-structured lipid-containing particles inside or outside cells [3]. Lipid metabolic defects can change lipid composition and/or distribution inside or outside a cell and may exert significant effects on cell physiology. This chapter will focus on possible effects of lipid metabolic defects on activities of voltage-gated ion channels at the molecular and cellular levels, which are probably part of the basis for organismal level phenotypes observed in human patients [4].

All cells have specific resting transmembrane electrostatic potentials, which require voltage-gated ion channels (VGICs), often a voltage-gated potassium (Kv) channel, as a feedback pathway, to be stabilized at certain levels under well-defined ionic conditions [5]. The transmembrane potential serves as a driving force or energy source for transporting various molecules and ions across the cell membranes. Inside a human body, all cells contain high K^+ and low Na^+ inside and face high Na^+ and low K^+ outside (an exception is the cochlear inner hair cells whose top ends meet with high-K^+ endolymph) [6]. The ionic differences are often maintained by Na^+-pump at the expense of ATP. Ion channels allow specific ions to flow down their electrochemical gradients and dissipate the chemical energy [5, 7]. For these channels to function well, lipid molecules in cell membranes must maintain stable ion gradients by separating the inside and outside milieus of a cell reliably with negligible shunting pathways, due to hydrophobicity in the membrane core and the high energy barriers for hydrated ions to cross. These channels therefore must reside in lipidic environments to be physiologically relevant. Their interactions with the lipid molecules are presumably a critical factor for shaping their functions [4, 8]. The same are expected to be true for other transmembrane proteins that have been evolved to function in specific lipid environments in their native niches.

VGICs belong to a superfamily of integral membrane proteins that have distinct voltage-sensor domains (VSDs) responsible for detecting changes in transmembrane electrostatic potential [5]. Except voltage-gated proton channels whose proton-conducting pores are within their VSDs, all known VGICs are made of four subunits / domains that together form a central pore surrounded by four VSDs in the periphery. For these channels, each subunit/domain contains a pore region and a VSD. Each pore region has two transmembrane helices and a pore loop, and each contributes to one quarter of an ion-conducting pore [9]. Each VSD contains four transmembrane segments (S1–S4) that are folded into a four-helix bundle, and harbors intrinsic positive charges, which are mostly carried in the extracellular half of the fourth segment (S4), physically relocate in response to change in transmembrane electric field and are called gating charges [10]. A VSD switches from a resting state to an activated state through sequential translocation of the fixed gating-charge-contributing residues (mainly in S4) from the intracellular side of the transmembrane electric field to the extracellular side [11–15]. The physical movement of typical four residues in 3 to 4 helical turns has been extensively studied. Presumably, five sequential states are required to account for moving all four positively-charged residues from the intracellular side to the extracellular side of the transmembrane electric field [16]. Crystallographic data have so far observed local rearrangements around one or two residues, making the general model of four-step motions reasonably well accepted [11]. However, this type of model may still be a simplification and may need modifications for a particular VSD [7]. We will not delve into such subtle details because a general mechanistic two-state model is often sufficient for our thermodynamic and kinetic considerations [7, 17].

From the angle of molecular composition of a bilayer membrane, lipid molecules in the two leaflets are arranged in juxtaposition to have hydrophobic tails from two leaflets face each other in the interior, achieve high stability and minimize trans-bilayer flipping of polar or charged lipid groups [7]. Laterally, lipids and proteins can form small functional units or domains, and lipid/protein composition may vary from one location to another in the plasma membrane of a cell. Across a cell membrane, the lipid composition of the two leaflets often differs from each other. The canonical fluid-mosaic model of a biomembrane suggests that a membrane protein complex and the few layers of lipid molecules surrounding it form a functional unit [1, 7, 8]. This same notion probably applies to the intracellular membranes as well, highlighting the importance of lipid-protein interactions in every biological membrane and the general proposal that a membrane protein evolves and becomes adapted to the lipid environment of its native niche in membrane [1].

Changes of lipid environments in the native niche of a membrane protein are thus expected, especially for those proteins whose conformational changes are particularly sensitive to lipids, to generate significant functional effects, which may alter cell physiology [17]. Lipid metabolic defects may cause changes in lipid uptake, degradation or synthesis of lipids, recycling and redistribution of lipids in plasma and organellar membranes, and/or removal of lipids from organelles or from cells through vesicles, protein binders or carriers, or enzymatic activities [6]. We will focus on the lipid metabolic effects on VGICs in the next sections, mainly because these channels have their VSDs at the lipid-protein interfaces and are particularly sensitive to their lipid environments [8, 18]. Our general theme is that the lipid-metabolic defects may change the so-called lipid-dependent gating of VGICs and bring about many of the neurological phenotypes or the functional and structural changes in organs or tissues outside the nervous systems [8].

2. Lipid metabolic defects and lipid distribution among cell membranes

There are at least five important processes for lipid metabolism and homeostasis (**Figure 1**) [19]: (a) lipid uptake and absorption/assimilation from diets, (b) *ab initio* synthesis of lipids from building blocks, (c) lipid recycling including transport from one organelle to another, release from lipid storage, conversion from one type to another by specific enzymes, (d) lipid degradation or removal from cells through exosomes or carrier protein complexes, and (e) storage of lipid molecules in tissues/organs, storage particles like lipid droplets or accumulation in lysosome-related compartments in cells. These processes are integrated and contribute to a complicated network in regulation of lipid homeostasis (**Figure 1**). We will not elaborate on these processes, but instead will focus on some of the key changes that will likely affect the voltage-gated ion channels [4].

Based on chemical structures and their arrangements in bilayers, all lipid molecules in human cell membranes fall into two groups — phospholipids and nonphospholipids, depending on whether a lipid has a phosphodiester group in its headgroup region [4] (**Figure 2**). Typically, phospholipids include glycerophospholipids, phosphosphingolipids and phosphono-sphingolipids (sphingomyelin as a prominent example), and others that contain a properly positioned phosphate group,

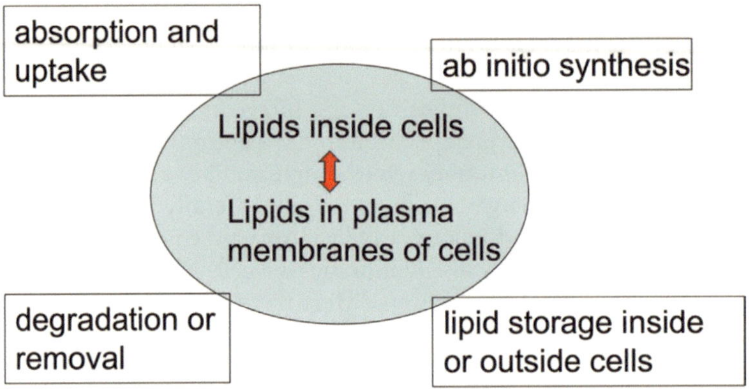

Figure 1.
Lipid homeostasis at systems and cellular levels. The lipids can be taken into or removed from the cells, can be synthesized inside the cells, can be stored insider or outside the cells, and can be recycled or degraded in cells.

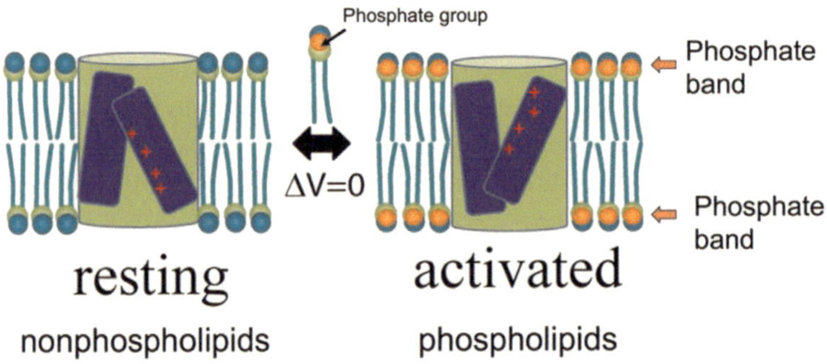

Figure 2.
Compositional difference between bilayers made of phospholipids and nonphospholipids contributes to lipid-dependent gating effects on a voltage-gated ion channel. The phosphate bands endow significant chemical properties that are unique to a phospholipid bilayer. A voltage-sensing domain (VSD) is favored energetically in a resting state when it is surrounded by nonphospholipids, and is stabilized in an activated state in the presence of surrounding phospholipids.

such as cholesterol-phosphate, ceramide-1-phosphate, geranyl phosphate, isopentenyl phosphate, etc. [20]. The phospho-diester groups in a typical phospholipid bilayer are aligned to form two bands that have high charge density and participate in extensive H-bonding networks, making a prominent chemical feature that is missing in a bilayer made of nonphospholipids. It remains unclear and untested whether a sulfate group can replace a phosphodiester group in the right locations of a lipid bilayer, even though its chemistry is different and in general it cannot substitute for a phosphate in biochemical settings. As a demonstration of the importance of the chemical bonding structures at the phosphate bands, plasmalogens deserve mention here [21, 22]. A plasmalogen contains one normal ester-bonded (at the *sn*-2 position of the glycerol backbone) acyl chain and one ether-linked chain (at the *sn*-1 position). They make 5–20% of the phospholipids in most mammalian cell membranes and account for ~18% of total phospholipid mass in humans [23]. Their vinyl-ether linkages are thought to scavenge reactive oxygen species (ROS). Their presence may introduce changes in biophysical properties of biological membranes, such as thickness, rigidity, fluidity, phase separation, fusion, etc. and may dictate their importance to myelin sheaths in nervous systems. Metabolic changes in plasmalogen abundance can alter CNS functions and have been proposed as an etiological factor of Alzheimer's disease (AD) [23].

The nonphospholipids include a more diversified group, including members of fatty acids, glycerolipids, sphingolipids, sterol lipids (cholesterol and analogs), prenol lipids, saccharolipids and polyketides [20]. Among them, cholesterol (CHOL or Chol) and glycosphingolipids (such as gangliosides) are quite abundant in human cells, especially in cells of CNS. For example, a human adult brain accounts for $\sim$3% of body weight, but contains nearly 25% of total body cholesterol [24]. All cholesterol in the brain is synthesized *de novo* in astrocytes. Cholesterol is not only a structural component in the branched membrane systems of a neuronal cell in the brain, but also serves as a precursor for critical signaling molecules and hormones in the system [19]. Not surprisingly, metabolic changes in the abundance or distribution of cholesterol, or nonphospholipids in general, is important to the proper functions of the CNS [25].

Metabolic defects in fatty acids, phospholipids or nonphospholipids often lead to changes in their relative abundance in circulation, tissues and organs, or in plasma membranes and intracellular membranes of individual cells, affecting both their structural roles, transport, distribution, recycling, and signaling functions [25–34]. Many genetic defects may be lethal in early stages of life. For examples, the most prominent fatty acid oxidation disorder, medium-chain acyl-CoA dehydrogenase deficiency (MCADD), may cause dehydration, lethargy and hypoketotic hypoglycemia, and result in early onset liver dysfunction and brain edema, and thus a high mortality rate [27]. Phospholipid metabolic defects, such as the ones leading to high or low phosphatidylcholine (PC)/phosphatidyl-ethanolamine (PE) ratios, can affect the formation and stability of triacylglyceride (TAG)-rich lipoproteins and alter *de novo* lipid synthesis via sterol regulatory element binding protein (SREBP) pathway. Changes in PC/PE ratios are linked to liver failure, impaired liver regeneration, etc., and genetic mutations affecting PC/PE ratios may give rise to embryonic lethality [29]. Similarly, multiple monogenic defects in metabolism of non-fatty-acid nonphospholipids alter their distribution in cells. The level of severity of their phenotypes varies, and can still lead to obvious functional changes even when organ development and cell growth appear normal in infancy (**Table 1**). Notably, altered nonphospholipid abundance in cell membranes almost always leads to changes in excitable cells, and invariably is manifested in severe or life-threatening neurological phenotypes (**Table 1**).

Based on these phenotypes, our central hypothesis is that these metabolic changes in the relative ratios of phospholipids and nonphospholipids in plasma membranes may change the activities of VGICs, and lead to phenotypical changes in nervous systems because of high abundance of these channels [4, 8, 18]. We will elaborate such connections in the next section first.

3. Lipid-dependent gating of voltage-gated ion channels (VGICs)

From the fluid mosaic model of a biomembrane, we proposed that a VGIC and a few layers of lipid molecules (called annular lipids here; **Figure 2**) right next to its transmembrane footprint together form a functional unit [1, 35]. We usually call the control of channel opening and closing as a *gating* process. Besides other factors, the gating energetics and physiological functions of a VGIC is also a function of lipid composition in the annular layer. As listed in **Table 1**, the metabolic defects of

Lipid types	Diseases	Mutated genes	Affected lipids	Impacted organs	Prevalence rate
Fatty Acids	medium-chain acyl-CoA dehydrogenase deficiency (MCADD) [28, 29]	ACADM	elevated acylcarnitine (C8, C10, C10:1)	liver, brain	1:20,000 in Northern European Caucasian
	X-linked adrenoleuko-dystrophy (X-ALD) [30]	ABCD1 and down-regulated ALDP	very long-chain fatty acids (VLCFAs)	Nervous systems and adrenal glands	1 in 20,000 to 50,000 worldwide
Phospholipids	PC synthesis deficiency [31]	CHKA / CHKB	phosphatidylcholine (P-choline)	global (embryonic lethal)	rare
	PE deficiency [31]	EKI2	P-ethanolamine	global (neonatal lethal)	rare
Non-phospholipids (other than fatty acids)	Cerebrotendinous Xanthomatosis (CTX) [25, 32]	CYP27A1	p450 oxidase; bile acids and lipids; sterol metabolites	brain (ataxia, dystonia, epilepsy, etc.)	< 5 in 100,000 worldwide; R362C 1 per 50,000 Caucasians
	Fabry's disease [33]	GLA	Glycolipids	Skin, heart, kidney, brain	1 in 40,000 to 60,000 male, less in female
	Niemann-Pick diseases type A, B, C [21, 34].	ASMD (type A, B); NPC1 and NPC2	Sphingomyelin, cholesterol	Tissue, liver, spleen, lymph nodes, brain, etc.	very rare, 500 cases diagnosed worldwide
	Tay-Sachs disease [22]	HEXA	GM2 gangliosides	Nervous system, muscle, etc.	1 in 90,000; mainly in infants, progressive and fatal.

Table 1.
Examples of metabolic defects in fatty acids, phospholipids and other nonphospholipids.

nonphospholipids (including fatty acids) may change the distribution and redistribution of these lipids in native cell membranes, whose effects may penetrate the annular layers of a VGIC that happens to be in the region and alter its functions. Altered VGIC activities may change the excitability of neurons and muscle cells in specific neural circuits or in muscles, leading to severe pathophysiological effects at tissues, organs, or organismal levels. To keep focused, we will discuss changes in relative abundance of nonphospholipids and phospholipids in plasma membranes where the VGICs contribute to key physiological events. We will not discuss the effects of lipid metabolic defects on the VGICs in intracellular membranes because of insufficient studies of their connections to lipid metabolic diseases so far [36].

Our studies in lipid-dependent gating of Kv channels led us to a more general concept of nonphospholipid-mediated gating effects on VGICs (**Figure 3**) [4, 17]. A typical depolarization-activated Kv channel undergoes transitions between different states in a voltage-dependent or -independent manner (**Figure 3A**) [37, 38].

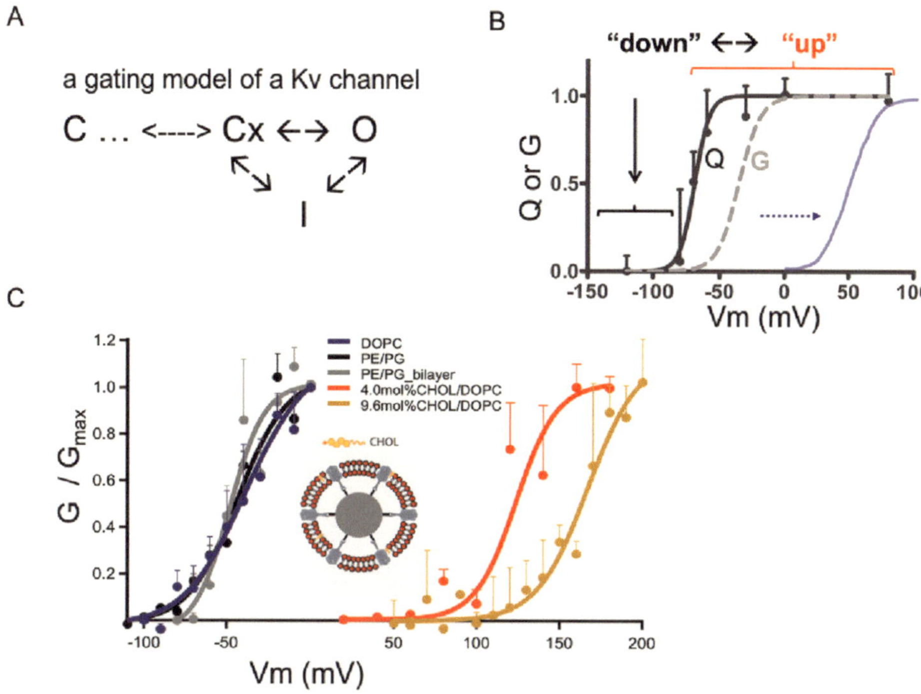

Figure 3.
Lipid-dependent gating directly alters properties of a VGIC. (A) A typical gating model for a VGIC that is depolarization-activated. C: closed state(s); Cx: closed state right before opening; O: open state; and I: inactivated state; (B) Changes of gating charge (Q) and conductance (G) as a function of transmembrane potential (Vm). The VSD switches from a resting ("down") to an activated ("Up") state, while the channel pore changes from the closed to the open state. The predicted change of the G-V or Q-V curve in the presence of nonphospholipids is a right shift (purple arrow and trace) with the same or a shallower slope; and (C) The measured changes of G-V curves of the KvAP channel in the presence of different lipids in bead-supported unilamellar membranes (bSUMs) made of different lipids (inserted cartoon). Those on the far left were from phospholipid membranes, while the red and orange curves were in DOPC bilayers doped with small amounts of cholesterol. From references [8, 18] with permissions following PubMed open access policy.

The measured gating charge (*Q*) movement (due to changes in the voltage sensor domain) or conductance (*G*) change (due to the opening of the pore domain) can be presented as a function of transmembrane potential (*Q-V* or *G-V* in **Figure 3B**). From the general concepts in **Figure 2**, the nonphospholipids favor the resting (closed) state of a VGIC. A prediction is that the *Q-V* and/or *G-V* would shift to the right and may keep the same slope or become shallower due to procrastinated conformational changes in certain steps (purple trace in **Figure 3B**) [39, 40]. Indeed, when we constructed a bead-supported unilamellar membrane (bSUM) with precisely controlled lipid composition and channel orientation in bilayers and without any additional solvents (e.g., no decane), it was feasible to measure accurately the shifts of *G-V* curves in the presence of 4.0% or 9.6% cholesterol (**Figure 3C**) [41]. Due to decane islands and annuli in planar lipid bilayers and potential (or preferential) interaction (packing) of cholesterol with decane, it was difficult to control cholesterol concentration and phase behavior of planar bilayers when bilayer experiments were performed by introducing cholesterol or glycolipids [42]. The bSUMs were thus better when planer glass electrodes were utilized for electrophysiological recordings from channels in them, and so far, remain the only system that is suitable to control lipid composition and channel orientation at the same time. Our results agree very well with the predictions of lipid dependent gating (**Figure 3C** vs. **B**).

Similar physiological changes have been reported for Nav1.9 channels in the sensory neurons and Kv3 channels in acutely isolated dopaminergic neurons [43, 44], suggesting a more generic nature of the nonphospholipid-dependent gating effects on voltage-gated ion channels. When more experimental data published in literature were studied, a generally agreed conclusion is that loading cell membranes by methyl-beta-cyclodextrin-cholesterol causes consistent right shifts of all VGICs (Kv, Nav and Cav channels) in different cells [45–56], whereas divergent or conflicting data were reported after cholesterol depletion by methyl-beta-cyclodextrin [17]. We know that strong depletion may cause significant changes to membrane integrity and submembrane structures inside cells. It is therefore reasonable to put more trust in the cholesterol-loading data and be very cautious of the results from cholesterol depletion in cell membranes. It is hence reasonable for us to conclude that the published data in literature agree with the predictions in **Figure 3B** and our hypothesis in **Figure 2**.

4. Energetics behind lipid-dependent gating

Thermodynamically, the free energy, $\Delta G\ (n, T, V, P, Q, Vm, \mu_P, \mu_S, \mu_L)$, for a VGIC in a lipid membrane can be treated as a function of molecular quantities (n), temperature (T), volume (V), pressure (P), charge (Q), transmembrane electrostatic potential (Vm)and chemical potentials of the protein (μ_P), solutes/solvents (μ_S), and lipids (μ_L) [4, 8, 17]. The charge movement across a gradient of electrostatic potential leads to a free energy change of $Q\ {}^*\Delta Vm$ or its integral over time. The *G-V* curves did not change steepness significantly when cholesterol was added, suggesting a similar or the same gating model with similar apparent gating charges ($Q \sim 2.0$–$3.0\ e_0$). A right shift by 200 mV (as in **Figure 3C**) is equivalent to a minimal energy change of 9.2 *kcal/mol* in stabilizing the closed states, slightly more than the hydrolytic energy of one ATP.

For VGICs, the chemical potential changes from the lipid bindings are manifested indirectly through energetic compensation from the gating charge movement. If we consider that an average number of lipid-binding sites is $<m>$, and there is an average change in affinity of the binding sites. If we use $k_{d,1}$ and $k_{d,2}$ to represent geometrically averaged apparent dissociation constants of lipids at equivalent binding sites in the closed and open states, respectively and consider no mutual interference between them, a simplified expression of free energy change from the lipid-binding sites is - $<m> RT\,ln\,(k_{d,1}/k_{d,2})$. A typical VGIC allows ~120 lipid-binding sites in the first layer of annular lipids around it. To make for the estimated free energy change from G-C curves, that is ~9.2 *kcal/mol*, ~10 lipid-binding sites are sufficient if each on average contributes ~1.5 *RT*, which equates to a $k_{d,2}/k_{d,1} \sim 4.5$. A small change in relative affinity of a group of binding sites hence suffices to cause a major shift in *G-V* (and *Q-V*) of a VGIC. In other words, the changes in affinity can happen to those of low, intermediate or high-affinities, but are not limited to the high-affinity binding sites because the latter usually play a critical structural role in stabilizing the 3D structure of a functional membrane protein.

The lipid-dependent gating is expected to impact on both Nav and Kv channels. For a neuron that fires action potentials (APs) using the Nav and Kv channels, the nonphospholipid-induced gating effect will make it more difficult to open Nav channels, which may result in a larger number of Nav channels needed to allow the cell to reach the threshold for firing action potential. Meanwhile, the lipid-dependent gating of Kv channels will slow down the repolarization of the transmembrane potential.

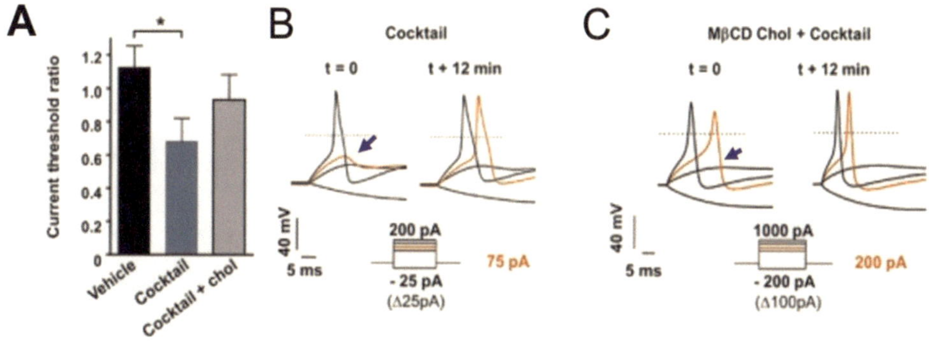

Figure 4.
Physiological effects of cholesterol-dependent gating of Nav1.9 channels in dorsal root ganglion (DRG) cells. (A) Carrageenan cocktail decreases the threshold for AP firing in DRG neurons, concurrent with decreased cholesterol content in the cells. MβCD/Chol (20 mM) treatment decreased the effects; (B) Control neurons had a lower threshold for AP firing, and only 75 pA current injection was needed to elicit an AP (blue arrow; t = 0 in B); and (C) Neurons pre-treated with MβCD/Chol for 10 minutes showed much higher threshold for AP firing (200 pA; blue arrow), due to cholesterol-dependent gating effects on Nav1.9 channels. In both cases the cocktail treatment (t + 12 min) increased the AP firing frequencies (but more pronounced in B than in C), and shortened the AP duration (especially in C). Adapted from reference [43] with permissions under the free access policy of EMBO press.

The net effects may include a shift in resting membrane potential, a shift in the threshold for AP firing, a decrease in the AP firing frequency, elongation of the AP duration, or a decrease in AP overshoots, etc. Consistent with these predictions (to the opposite direction due to cholesterol decrease), in cultured dorsal root ganglion (DRG) neurons, cholesterol-dependent gating effects on the Nav1.9 channels were found to be the root cause when local cholesterol decrease in the neurons resulted from carrageenan injection caused hyperexcitability and enhanced pain sensation. As an example of the physiological data, in **Figure 4A**, decrease of thresholds happened when cholesterol content was decreased and AP firing frequencies were increased after carrageenan cocktail treatment (**Figure 4B** and **C** at t + 12 min). Notably, in **Figure 4B** and **C** (t = 0 min), the neurons preloaded with cholesterol (MβCD/Chol 20 mM) showed an AP threshold for current injection of 200 pA, much higher than the control 75 pA, consistent with the predictions of cholesterol-dependent gating.

Similar to what was reported by the Padilla group, my group in collaboration with Dr. Yuqing Li's lab at the University of Florida raised an NPC1-I1061T-knockin model mouse [57] and discovered that the AP firing frequency in triphasic cerebellar Purkinje neurons was ~30% lower in the mutant than that in the wild-type (QXJ, unpublished observations). The decreased cholesterol-content likely activated Kv channel more significantly than the Nav or Cav channels in the Purkinje neurons, leading to a marked inhibitory effect. More detailed studies are needed to examine the channels in these Purkinje neurons to define the lipid dependent gating effects better.

5. Computational methods for low-to-intermediate-affinity binding sites of non-phospholipids on membrane proteins

Our theoretical analysis in the last section suggested a likely scenario for the lipid-dependent gating to function in a cell membrane, which contains abundant content of nonphospholipids. It relies on changes of ten or so low- to-intermediate-affinity binding sites because high lipid content would saturate all high-affinity binding sites

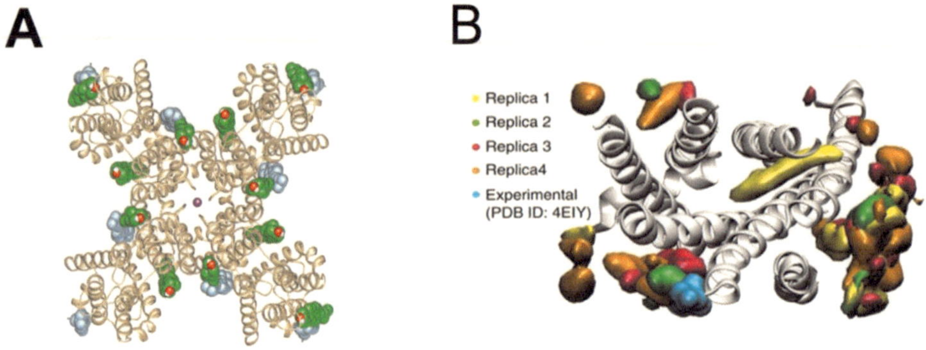

Figure 5.
Computational analysis of action sites for lipid-dependent gating. (A) Computational docking suggested multiple cholesterol-binding sites in the intra- (green balls) and extra-cellular (blue) leaflets around a Kv1.2 (PDB: 2R9R) structure, leading to 20 possible sites. The channel is viewed from the extracellular side. Adapted from reference [58] with permissions following the PubMed open access policy; and (B) Cholesterol-binding densities (replicas 1–4) around an A2AR (PDB: 4EIY; only helices showed in gray) predicted from molecular dynamics simulation, indicating at least 8 binding sites. The cholesterol resolved in the crystal structure is in cyan and the structure is viewed from the extracellular side. Adapted from reference [59] with permissions following the PubMed open access policy.

continuously. However, it is difficult to study these non-structural lipids because of possible heterogeneity in the bound poses and/or slightly larger volumes of such binding sites to accommodate different poses, making these almost impossible to pin down accurately using conventional structural methods.

Computational analysis has been used to make some predictions that may be tested experimentally (for examples, **Figure 5**) [7]. For cholesterol, the consensus binding motifs predicted in the past carried uncertainties. Docking and energy minimization was used to identify 20 possible binding sites (limited to 20; **Figure 5A**), eight at the periphery and 12 near the interfaces between the pore domain and the VSDs. The same analysis of Kv7.1 led to 16 binding sites to the periphery of four VSDs and four at the pore-VSD interfaces. Experimental testing of these sites is still lacking.

In a similar situation, structural studies of G-protein coupled receptors (GPCRs) revealed only limited number of cholesterol-binding sites, but it is known that cholesterol is critically important for physiological functions of various GPCRs. Bovine rhodopsin, for example, encounters a high cholesterol content at an 8:1 molar ratio that are important for ensuring high stability of the protein in membrane [60]. Molecular dynamics (MD) simulation in GPCRs has predicted multiple binding sites (**Figure 5B**) that were not detected by structural studies and are not usually agreed on among different proteins. A majority of the predicted binding sites are likely not conserved and may not be well occupied for structural identification [61]. MD simulation may uncover more cholesterol binding events right next to each receptor [59, 62] (**Figure 2**). These studies highlight the difference between structural lipids of high binding affinity and those loose and dynamic binding sites with lower affinity but may be the major contributors to the lipid-dependent gating.

6. Better technologies desired for studying lipid-dependent gating

Technically, we are rather limited in studying lipid-dependent gating even in a small patch of cell membranes because of complicated lipid environments in face of phase separation and domain organization, lipid heterogeneity and asymmetry in two leaflets, membrane-matrix or membrane-cytoskeleton interactions, etc. The dynamic occupancy of the low-to-intermediate affinity lipid-binding sites, which were predicted to be chiefly responsible for lipid-dependent gating, becomes a roadblock against the structural / biophysical studies of their effects on different VGICs in both reconstituted and native cell membranes. Ideally, we would like to use high-resolution real-time methods to analyze the lipid composition or distribution within the annulus or annuli of one or a few VGICs that are in a uniform lipid environment. With available technologies, single molecule detection of nonphospholipids around a VGIC with sufficient spatial resolution is almost impossible in the presence of a sea of lipids in a biomembrane. In a recently-published book chapter [1], QXJ discussed the possible development of single molecule (ion) detection in mass spectrometry, surface-enhanced Raman spectrometry [63], nanoscale secondary ion mass spectrometry (NanoSIMS) [64], and high-resolution atomic force microscopy (AFM). These methods will need significant developments to achieve sufficient resolutions and differentiate nonphospholipids from phospholipids around individual VGICs in membranes. Here we will describe a few directions that may practically enrich our understandings of lipid-dependent gating in the near future.

First, even though the binding poses and positions of lipids as predicted from computational analysis often carry a lot of uncertainty, many of them can be tested in functional studies in order to identify a prominent one or a few that are functionally relevant and allow testable predictions to be formulated. The sites around the Kv channels (such as those in **Figure 5A**) may be suitable targets. Alternatively, zero-length cross-linkers can be introduced and pin down the low affinity cholesterol-binding sites around a VGIC [65]. These sites can be further verified by functional studies. Single molecular fluorescence resonance energy transfer (smFRET) may be another method that is suitable for probing dynamic lipid-protein interactions within the annular layer of a VGIC when suitable probes can be introduced into the proteins to serve as sensors.

Second, genetic methods may be suitable to probe the co-evolution of VGICs in contrastingly different lipid environments. For example, it is reported that some strains of bacteria isolated from Lake Matano with phosphate concentration < 50 nM, can grow without added phosphorus, and replace nearly 100% of membrane lipids with amino- or glycolipids, like what was reported for photoautotrophs and heterotrophs [66]. Because these strains have evolved to survive in a phosphorus-limited native niche, they can make a good model system to assess how the VGICs have evolved to deal with the selection pressure. The amino-glycolipids in the acidic environment in these bacteria must be positively charged so that a VGIC would face a very different environment for voltage-gating than a phospholipid-rich membrane.

Third, the potential of cryo-electron tomography has been gradually improved for high-resolution imaging [67]. The major bottleneck still resides with the limited dose of electrons used per exposure. For unilamellar biomembranes that are of 4–6 nm in thickness, it might be feasible to introduce graphene films to both sides so that the individual membranes can tolerate a much higher dose of electrons for imaging [68, 69]. If the dose rate can improve by 10–100 folds, the tomographic resolution will be augmented significantly. If so, it might become feasible to locate individual lipid molecules directly or achieve the minimum to distinguish phospholipids from nonphospholipids within the annular layer of a particular VGIC.

Fourth, from a physiological angle, it is interesting to ask if targeted delivery of nonphospholipid analogs may be used for treating the life-threatening neurological diseases suffered by human patients carrying lipid metabolic defects [70, 71].

Biopsy fibroblasts from these patients can be isolated and reprogrammed into neurons in order to test these analogs before their clinical usage will be pursued. These analogs may be easier to deliver and can be constructed as biodegradable and will satisfy current shortage of suitable clinical measures for these patients and extend their lifespans for better quality of lives [72].

Fifth, super-resolution fluorescence microscopy (srFM) that can reach 1.0 nm or better in resolution may be suitable for imaging nonphospholipids directly if proper labeling can be done. For example, MINFLUX nanoscopy shows the promise to localize individual fluorophores with ~1.0 nm resolution [73]. With a fluorophore on a VGIC and labels on the nonphospholipids, a system may be constructed for determining the distance of nearest nonphospholipids to the channel. If the tetrameric nature of the channel can be used to orient it in membrane, it may even be feasible to separate the nonphospholipids at the periphery of the VSDs from those closer to the pore domain (as in **Figure 5A**). Proper labels still await development.

It is for sure a big challenge, if not an impossible task, to exhaust feasible avenues for the near future. The exciting perspective is that with new methods, it will for the first time become feasible to probe the importance of annular lipids to membrane protein functions with experimental evidence, far better than the qualitative predictions from a macroscopic view of the fluid mosaic model [35].

7. Conclusions

Lipid-dependent gating of VGICs and metabolic defects in fatty acids and other categories of nonphospholipids are probably tightly connected. The underlying natural driving force appears the co-evolution of voltage-gated ion channels with the phospholipid-rich biomembranes in all three-kingdoms of life. The thermodynamic principles are hinged on the collective interactions at a dozen or so lipid-binding sites of low-to-intermediate affinity with varying occupancy. New methods are still desired to directly dissect the annular lipids around individual channels for a physical test of the lipid-dependent gating hypothesis.

Acknowledgements

We would like to thank Dr. Roderick MacKinnon for supporting QXJ to pursue the direction of lipid-dependent gating as an independent investigator. The work in the Jiang laboratory in lipid-dependent gating and other programs has been supported by different funding agencies, including NIH (R21GM131231, R01GM111367, R01GM093271, & R01GM088745), AHA (12IRG9400019), CF Foundation (JIANG15G0), Welch Foundation (I-1684), CPRIT (RP120474) and intramural funds at the UT Southwestern Medical Center, the University of Florida and the Hauptman-Woodward Medical Research Institute (HWI). We should also mention that the earlier efforts (before year 2014) of using electron crystallography to elucidate the structural basis of lipid-dependent effects on Kv channels were mainly invested by the Tamir Gonen lab because of limited access of high-end cryo-EM and lack of expertise in analyzing 2D crystals in the Jiang lab. We thank Dr. Gonen and Dr. Liang Shi for transferring the 2D crystal project to the Gonen lab and investing their time in crystal optimization, data collection and analysis in face of technical difficulties and pitfalls in different areas. Only till recently (after year 2019), the Jiang lab was able to gain enough cryo-EM time in order to achieve near-atomic resolutions of Kv channels in lipid environments. We thank Dr. Gaya Yadav, now at TAMU, for his persistent efforts in this area.

Conflict of interest

The authors declares no conflict of interest.

Notes

None.

Appendices and nomenclature

None.

Author details

Qiu-Xing Jiang[1,2,3]* and Felix Chin[1,4]

1 Department of Physiology and Biophysics, University at Buffalo, Buffalo, NY, USA

2 Department of Medicinal Chemistry, University of Florida, Gainesville, FL, USA

3 Laboratory of Molecular Physiology and Biophysics, Hauptman-Woodward Medical Research Institute, Buffalo, NY, USA

4 Perelman School of Medicine, University of Pennsylvania, Philadelphia

*Address all correspondence to: qxjiang@ufl.edu; qiuxingj@buffalo.edu or jiangq9992003@gmail.com

References

[1] Jiang QX. Introudction: New tools for challenges in membrane biology. In: Jiang QX, editor. New Techniques for Studying Biomembranes. Methods in Signal Transduction. 1st ed. Boca Raton, FL: CRC Press; 2020. pp. 1-8

[2] Wigge C, Stefanovic A, Radjainia M. The rapidly evolving role of cryo-EM in drug design. Drug Discovery Today: Technologies. 2021. pp. 91-102

[3] Alberts B, Johnson A, Lewis J, Raff M, Roberts K, Walter P. Molecular Biology of the Cell. 5th ed. New York: Galand Science; 2007

[4] Jiang QX. Lipid-dependent gating of ion channels. In: Catala A, editor. Protein-Lipid Interactions: Perspectives, Techniques and Challenges. U.S.: Nova Science Publishers, Inc.; 2018. p. 196

[5] Hille B. Ion Channels of Excitable Membranes. 3rd ed. Sunderland, MA, USA: Sinauer Associates, Inc.; 2001

[6] Iwasa J, Marshall W. Karp's Cell and Molecular Biology. 8th ed. USA: Wiley; 2016. p. 740

[7] Jiang QX. High-resolution structures of K(+) channels. Handbook of Experimental Pharmacology. 2021;**267**: 51-81

[8] Jiang QX. Cholesterol-dependent gating effects on ion channels. Advances in Experimental Medicine and Biology. 2019;**1115**:167-190

[9] MacKinnon R. Potassium channels. FEBS Letters. 2003;**555**(1):62-65

[10] Sigworth FJ. Voltage gating of ion channels. Quarterly Reviews of Biophysics. 1994;**27**(1):1-40

[11] Guo J, Zeng W, Chen Q, Lee C, Chen L, Yang Y, et al. Structure of the voltage-gated two-pore channel TPC1 from Arabidopsis thaliana. Nature. 2016; **531**(7593):196-201

[12] Kintzer AF, Stroud RM. Structure, inhibition and regulation of two-pore channel TPC1 from Arabidopsis thaliana. Nature. 2016;**531**(7593):258-262

[13] Kintzer AF, Stroud RM. On the structure and mechanism of two-pore channels. The FEBS Journal. 2018; **285**(2):233-243

[14] Payandeh J, Scheuer T, Zheng N, Catterall WA. The crystal structure of a voltage-gated sodium channel. Nature. 2011;**475**(7356):353-358

[15] Payandeh J, Gamal El-Din TM, Scheuer T, Zheng N, Catterall WA. Crystal structure of a voltage-gated sodium channel in two potentially inactivated states. Nature. 2012; **486**(7401):135-139

[16] Tao X, Lee A, Limapichat W, Dougherty DA, MacKinnon R. A gating charge transfer center in voltage sensors. Science. 2010;**328**:67-73

[17] JIang QX, Levitan I. Principles of cholesterol regulation of ion channels. In: Dopico AN, Ba AM, editors. Cholesterol: From Chemistry and Biophysics to the Clinic. London, San Diego, Oxford, Cambridge: Elsevier Inc.; 2022. pp. 1021-1037

[18] Zheng H, Liu W, Anderson LY, Jiang QX. Lipid-dependent gating of a voltage-gated potassium channel. Nature Communications. 2012;**2**:250

[19] Berg JM, Tymoczko JL, Gotto JG, Stryer L. Biochemistry. 9th ed. USA: Macmillan; 2019

[20] MAPS® L. The LIPID MAPS® Lipidomics Gatway 2022. Available from: https://www.lipidmaps.org/

[21] Su XQ, Wang J, Sinclair AJ. Plasmalogens and Alzheimer's disease: A review. Lipids in Health and Disease. 2019;**18**(1):100

[22] Bichenkov E, Ellingson JS. Temporal and quantitative expression of the myelin-associated lipids, ethanolamine plasmalogen, galactocerebroside, and sulfatide, in the differentiating CG-4 glial cell line. Neurochemical Research. 1999;**24**(12):1549-1556

[23] Wood PL, Khan MA, Mankidy R, Smith T, Goodenowe DB. Plasmalogen deficit: A new and testable hypothesis for the etiology of Alzheimer's disease. In: SDL M, editor. Alzheimer's Disease Pathogenesis - Core Concepts, Shifting Paradigms and Therapeutic Targets. 2011

[24] Huang X, Sterling NW, Du G, Sun D, Stetter C, Kong L, et al. Brain cholesterol metabolism and Parkinson's disease. Movement Disorders. 2019; **34**(3):386-395

[25] Cariati I, Masuelli L, Bei R, Tancredi V, Frank C, D'Arcangelo G. Neurodegeneration in Niemann-pick type C disease: An updated review on pharmacological and non-pharmacological approaches to counteract brain and cognitive impairment. International Journal of Molecular Sciences. 2021;**22**(12)

[26] Shekhawat PS, Matern D, Strauss AW. Fetal fatty acid oxidation disorders, their effect on maternal health and neonatal outcome: Impact of expanded newborn screening on their diagnosis and management. Pediatric Research. 2005;**57**(5 Pt 2): 78R-86R

[27] Merritt JL, Norris M, Kanungo S. Fatty acid oxidation disorders. Annals of Translational Medicine. 2018; **6**(24):473

[28] Weber FD, Wiesinger C, Forss-Petter S, Regelsberger G, Einwich A, Weber WH, et al. X-linked adrenoleukodystrophy: Very long-chain fatty acid metabolism is severely impaired in monocytes but not in lymphocytes. Human Molecular Genetics. 2014;**23**(10):2542-2550

[29] van der Veen JN, Kennelly JP, Wan S, Vance JE, Vance DE, Jacobs RL. The critical role of phosphatidylcholine and phosphatidylethanolamine metabolism in health and disease. Biochimica et Biophysica Acta—Biomembranes. 2017;**1859**(9 Pt B): 1558-1572

[30] Hoflinger P, Hauser S, Yutuc E, Hengel H, Griffiths L, Radelfahr F, et al. Metabolic profiling in serum, cerebrospinal fluid, and brain of patients with cerebrotendinous xanthomatosis. Journal of Lipid Research. 2021;**62**: 100078

[31] Nie S, Chen G, Cao X, Zhang Y. Cerebrotendinous xanthomatosis: A comprehensive review of pathogenesis, clinical manifestations, diagnosis, and management. Orphanet Journal of Rare Diseases. 2014;**9**:179

[32] Azevedo O, Cordeiro F, Gago MF, Miltenberger-Miltenyi G, Ferreira C, Sousa N, et al. Fabry disease and the heart: A comprehensive review. International Journal of Molecular Sciences. 2021;**22**(9)

[33] Schuchman EH, Desnick RJ. Types a and B Niemann-pick disease. Molecular Genetics and Metabolism. 2017;**120**(1–2):27-33

[34] Cachon-Gonzalez MB, Zaccariotto E, Cox TM. Genetics and therapies for GM2 Gangliosidosis. Current Gene Therapy. 2018;**18**(2): 68-89

[35] Nicolson GL. The fluid-mosaic model of membrane structure: Still relevant to understanding the structure, function and dynamics of biological membranes after more than 40 years. Biochimica et Biophysica Acta. 2014; **1838**(6):1451-1466

[36] Cang C, Zhou Y, Navarro B, Seo YJ, Aranda K, Shi L, et al. mTOR regulates lysosomal ATP-sensitive two-pore Na(+) channels to adapt to metabolic state. Cell. 2013;**152**(4):778-790

[37] Schoppa NE, Sigworth FJ. Activation of shaker potassium channels. III. An activation gating model for wild-type and V2 mutant channels. The Journal of General Physiology. 1998;**111**(2): 313-342

[38] Islas LD, Sigworth FJ. Electrostatics and the gating pore of shaker potassium channels. The Journal of General Physiology. 2001;**117**(1):69-89

[39] Schoppa NE, Sigworth FJ. Activation of shaker potassium channels. II. Kinetics of the V2 mutant channel. The Journal of General Physiology. 1998; **111**(2):295-311

[40] Schmidt D, Jiang QX, MacKinnon R. Phospholipids and the origin of cationic gating charges in voltage sensors. Nature. 2006;**444**(7120):775-779

[41] Zheng H, Lee S, Llaguno MC, Jiang QX. bSUM: A bead-supported unilamellar membrane system facilitating unidirectional insertion of membrane proteins into giant vesicles. The Journal of General Physiology. 2016; **147**(1):77-93

[42] Lee S, Zheng H, Shi L, Jiang QX. Reconstitution of a Kv channel into lipid membranes for structural and functional studies. Journal of Visualized Experiments. 2013;77:e50436

[43] Amsalem M, Poilbout C, Ferracci G, Delmas P, Padilla F. Membrane cholesterol depletion as a trigger of Nav1.9 channel-mediated inflammatory pain. The EMBO Journal. 2018;**37**(8): e97349

[44] Gantz SC, Bean BP. Cell-autonomous excitation of midbrain dopamine neurons by endocannabinoid-dependent lipid signaling. Neuron. 2017; **93**(6):1375-87 e2

[45] Balajthy A, Somodi S, Petho Z, Peter M, Varga Z, Szabo GP, et al. 7DHC-induced changes of Kv1.3 operation contributes to modified T cell function in Smith-Lemli-Opitz syndrome. Pflügers Archiv. 2016; **468**(8):1403-1418

[46] Rudakova E, Wagner M, Frank M, Volk T. Localization of Kv4.2 and KChIP2 in lipid rafts and modulation of outward K+ currents by membrane cholesterol content in rat left ventricular myocytes. Pflügers Archiv. 2015;**467**(2): 299-309

[47] Heaps CL, Bowles DK, Sturek M, Laughlin MH, Parker JL. Enhanced L-type Ca2+ channel current density in coronary smooth muscle of exercise-trained pigs is compensated to limit myoplasmic free Ca2+ accumulation. The Journal of Physiology. 2000;**528** (Pt 3):435-445

[48] Xia F, Gao X, Kwan E, Lam PP, Chan L, Sy K, et al. Disruption of pancreatic beta-cell lipid rafts modifies Kv2.1 channel gating and insulin exocytosis. The Journal of Biological Chemistry. 2004;**279**(23):24685-24691

[49] Randich AM, Cuello LG, Wanderling SS, Perozo E. Biochemical and structural analysis of the hyperpolarization-activated K(+) channel MVP. Biochemistry. 2014; **53**(10):1627-1636

[50] Li Q, Wanderling S, Sompornpisut P, Perozo E. Structural basis of lipid-driven conformational transitions in the KvAP voltage-sensing domain. Nature Structural & Molecular Biology. 2014;**21**(2):160-166

[51] Ramu Y, Xu Y, Lu Z. Enzymatic activation of voltage-gated potassium channels. Nature. 2006;**442**(7103): 696-699

[52] Xu Y, Ramu Y, Lu Z. Removal of phospho-head groups of membrane lipids immobilizes voltage sensors of K+ channels. Nature. 2008;**451**(7180): 826-829

[53] Pouvreau S, Berthier C, Blaineau S, Amsellem J, Coronado R, Strube C. Membrane cholesterol modulates dihydropyridine receptor function in mice fetal skeletal muscle cells. The Journal of Physiology. 2004;**555**(Pt 2): 365-381

[54] Hajdu P, Varga Z, Pieri C, Panyi G, Gaspar R Jr. Cholesterol modifies the gating of Kv1.3 in human T lymphocytes. Pflügers Archiv. 2003;**445**(6):674-682

[55] Bowles DK, Heaps CL, Turk JR, Maddali KK, Price EM. Hypercholesterolemia inhibits L-type calcium current in coronary macro-, not microcirculation. Journal of Applied Physiology. 1985;**96**(6):2240-2248

[56] Chun YS, Shin S, Kim Y, Cho H, Park MK, Kim TW, et al. Cholesterol modulates ion channels via down-regulation of phosphatidylinositol 4,5-bisphosphate. Journal of Neurochemistry. 2010;**112**(5):1286-1294

[57] Praggastis M, Tortelli B, Zhang J, Fujiwara H, Sidhu R, Chacko A, et al. A murine Niemann-pick C1 I1061T knock-in model recapitulates the pathological features of the most prevalent human disease allele. The Journal of Neuroscience. 2015;**35**(21):8091-8106

[58] Lee AG. Interfacial binding sites for cholesterol on Kir, Kv, K2P, and related potassium channels. Biophysical Journal. 2020;**119**(1):35-47

[59] Guixa-Gonzalez R, Albasanz JL, Rodriguez-Espigares I, Pastor M, Sanz F, Marti-Solano M, et al. Membrane cholesterol access into a G-protein-coupled receptor. Nature Communications. 2017;**8**:14505

[60] Ruprecht JJ, Mielke T, Vogel R, Villa C, Schertler GF. Electron crystallography reveals the structure of metarhodopsin I. The EMBO Journal. 2004;**23**(18):3609-3620

[61] Prasanna X, Sengupta D, Chattopadhyay A. Cholesterol-dependent conformational plasticity in GPCR dimers. Scientific Reports. 2016;**6**: 31858

[62] Gimpl G. Interaction of G protein coupled receptors and cholesterol. Chemistry and Physics of Lipids. 2016; **199**:61-73

[63] Faried M, Suga K, Okamoto Y, Shameli K, Miyake M, Umakoshi H. Membrane surface-enhanced Raman spectroscopy for cholesterol-modified lipid systems: Effect of gold nanoparticle size. ACS Omega. 2019;**4**(9):13687-13695

[64] Gorman BL, Kraft ML. High-resolution secondary ion mass spectrometry analysis of cell

membranes. Analytical Chemistry. 2020; **92**(2):1645-1652

[65] Ferraro NA, Cascio M. Cross-linking-mass spectrometry studies of cholesterol interactions with human alpha1 glycine receptor. Analytical Chemistry. 2018;**90**(4):2508-2516

[66] Yao M, Elling FJ, Jones C, Nomosatryo S, Long CP, Crowe SA, et al. Heterotrophic bacteria from an extremely phosphate-poor lake have conditionally reduced phosphorus demand and utilize diverse sources of phosphorus. Environmental Microbiology. 2016;**18**(2):656-667

[67] Schur FK. Toward high-resolution in situ structural biology with cryo-electron tomography and subtomogram averaging. Current Opinion in Structural Biology. 2019;**58**:1-9

[68] Liu N, Zhang J, Chen Y, Liu C, Zhang X, Xu K, et al. Bioactive functionalized monolayer graphene for high-resolution Cryo-electron microscopy. Journal of the American Chemical Society. 2019;**141**(9): 4016-4025

[69] Llaguno MC, Xu H, Shi L, Huang N, Zhang H, Liu Q, et al. Chemically functionalized carbon films for single molecule imaging. Journal of Structural Biology. 2014;**185**(3):405-417

[70] Immordino ML, Dosio F, Cattel L. Stealth liposomes: Review of the basic science, rationale, and clinical applications, existing and potential. International Journal of Nanomedicine. 2006;**1**(3):297-315

[71] Rufino-Ramos D, Albuquerque PR, Carmona V, Perfeito R, Nobre RJ, Pereira de Almeida L. Extracellular vesicles: Novel promising delivery systems for therapy of brain diseases. Journal of Controlled Release. 2017;**262**: 247-258

[72] Marques ARA, Saftig P. Lysosomal storage disorders - challenges, concepts and avenues for therapy: Beyond rare diseases. Journal of Cell Science. 2019; **132**(2)

[73] Eilers Y, Ta H, Gwosch KC, Balzarotti F, Hell SW. MINFLUX monitors rapid molecular jumps with superior spatiotemporal resolution. Proceedings of the National Academy of Sciences of the United States of America. 2018;**115**(24):6117-6122

Chapter 10

Lipidomics as a Tool in the Diagnosis and Clinical Therapy

María Elizbeth Alvarez Sánchez, Erick Nolasco Ontiveros, Rodrigo Arreola, Adriana Montserrat Espinosa González, Ana María García Bores, Roberto Eduardo López Urrutia, Ignacio Peñalosa Castro, María del Socorro Sánchez Correa and Edgar Antonio Estrella Parra

Abstract

The lipids are essential compounds of cells, with biochemical and structural properties. Lipids are classified according to their chain length or saturation levels and biogenesis. Lipidomics is a spectroscopic and spectrometric technique, like Mass Spectrometry and Nuclear Magnetic Resonance, as well as bioinformatics to quantify and characterize the lipid profile. Lipidomics enables the fundamental understanding of lipid biology, the identification of drug targets for therapy, and the discovery of lipid biomarkers of disease cohorts. Therefore, lipidomics allows knowing the diagnosis and clinical follow-up in medical therapy towards any disease. In this way, the lipid profile allows us to monitor the administration of a clinical treatment and assertively diagnose human diseases.

Keywords: clinical biomarkers, lipid biomarker, lipidomic methodology, lipidomic profile, personalized medicine

1. Introduction

Lipids are prominent among the four main macromolecules (the others being amino acids, carbohydrates, and nucleic acids) in the diversity of molecular species [1]. They are essential to the biochemical and biophysical properties of all cells [2, 3]. Lipids serve as energy storage sources [4], homeostasis regulators [5], and nutrients; besides, they participate in events of pathophysiological importance [6]. Current estimates place the number of lipids at 100,000–500,000 [6].

Classification of lipids is based on their chain length, backbone, or saturation levels. Chain length classification is self-explanatory, while lipids are categorized into phospholipids, glycolipids, SPH, and sterols based on their backbone or saturated or unsaturated according to their saturation level [3, 7]. Complex

 IntechOpen

lipids, such as glycolipids and sphingolipids, are combinations of carbohydrates or sphingoid bases and fatty acids [3]. Ceramides are bioactive lipids that participate in biochemical reactions, such as metabolism, apoptosis, and inflammation, that produce cardiovascular diseases when dysregulated [8]. Lipids comprise around one-third of the components of cells and are essential to metabolism and intracellular signaling. Thus, fully understanding their dynamic is fundamental to preventing, diagnosing, and treating a wide range of human diseases [9], such as cancer, where lipid biosynthesis is often increased [10]. However, the diversity and number of lipids as well as their variation between individuals is a current topic under study [6].

Lipidomics is a sub-branch of metabolomics devoted to studying the complete lipid profile within a cell, tissue, or organism through spectroscopic and/or spectrometric methods in combination with bioinformatic analysis [4, 9, 11]. This combined approach provides a comprehensive understanding of the role of the lipid profile –the lipidome– in biological systems [12]. Thus, the stability and reproducibility of the analytical methods are critical for producing reliable output [13]. The degree of understanding of lipid biology available through lipidomics facilitates biomarker discovery and offers therapeutic possibilities [6]. A case in point, it developed a mouse model for the Gulf War Illness through lipidomics, which led to a therapy scheme that effectively modified the lives of roughly 250,000 soldiers affected by the disease. This chapter reports on the importance of lipidomics in diagnosing and clinical follow-up in medical therapy [14, 15].

2. Toward personalized medicine

2.1 Omics sciences: the big data age

Lipid analyses performed on clinical chemical analyzers, such as TAG, HDL-C, and FC, have traditionally been used to analyze and predict the state of health and/ or the risk of developing a disease [16]. But with the advent of the post-genomic era, these individual analyses are quickly being replaced by broad "-omics" studies, made possible by a rampant development of precision instrumentation and computational resources [4]. This technology allows analyzing a large number of biological samples in a profitable way for clinical tests [17]. Multi-omics research strategies have proven to be a comprehensive system for the identification of biomarker molecules, which can detect individuals with potential risk of diseases, as well as their nutritional evaluation [18]. Illnesses can be studied in greater detail through lipidomics since they comprehensively evaluate the metabolites and metabolic pathways [19], allowing the simultaneous study of several lipid species in body fluids [20]. This balance between flexibility and detail is the cornerstone of personalized medicine.

The use of omics over classical markers has clear advantages. It has, for instance, yielded more practical and valuable indicators for diabetes, contributing to a broader picture of the disease [21]. Moreover, lipidomics findings are being combined with other omics studies –such as transcriptomics, metabolomics, and proteomics–and advanced imaging to reach a previously unattainable degree of precision phenotyping in progressively numerous cohorts. Undoubtedly, biomarkers and pharmaceutical targets discovered through these combined approaches are rapidly gaining importance in clinical scenarios [22, 23].

2.2 Lipid biomarkers in the clinic

Biomarkers are quantifiable characteristics distinctive of metabolic or pathogenic processes; the latter enables diagnostic, treatment assignment, and response monitoring [18, 24]. Individual lipids and lipid profiles have been appointed as biomarkers of human diseases and are analyzed by means of various chromatography techniques coupled with mass spectrometry, as summarized in **Table 1**.

Pre-analysis	**Instrumentation**	**Analysis/Post-analysis**	**Ref.**
Disease (biological sample)	**(Mass analyzer) system**	**Results/Conclusions**	
44 cystic fibrosis patients (plasma)	(ion-trap) LC-MS	PC and LPC detected species	[25]
DM-2 and NAFLD patients (plasma)	GC-MS *NMR^{1}H	Liver fat content is strongly associated with Chol synthesis independently of obesity	[26]
6 **preeclampsia patients** (placental)	(Q-trap) LC-MS	Lipids of STBM are implicated in immune response, coagulation, oxidative stress and apoptosis.	[27]
7,500 Atrioventricular septal defect patients (blood)	(Q-Trap) LC-MS GC-MS	Biomarkers will be assessed for association, calibration, discrimination and reclassification	[23]
3 Gaucher disease patients (plasma and urine)	(ion trap) nLC–ESI–MS/MS	20 plasma and 10 urinary lipids were selected as significant species of Gaucher disease	[28]
10 CKD patients (stage 4/5 renal disease) (plasma)	(LTQ) LC-MS/MS	Lipid alterations in CKD disease (plasmenyl ethanolamines, sulfatides, Cer and Chol sulfate)	[29]
8 CVD patients: lipoprotein apheresis treatment (plasma)	LC-MS/MS	Increases of anti-inflammatory lipid mediators derived from AA or EPA and DHA.	[30]
150 coronary disorders Tunisian patients (blood)	GC/MS-SIM	PA and PS as biomarkers of peroxisomal metabolism disorders in atherosclerosis progression.	[31]
30 hypercholesterolemia pregnant women (plasma)	(QTOF) UPLC- MS/MS	PC (16:0/20:4) (18:0/20:4) lipid species in cord blood affected by gestational hyper-cholesterolemia.	[32]
75 anorexia nervosa patients (plasma)	GC/MS	Increased: ω-3 ALA, EPA. Decreased: ω-6 to ω-3, LA, ALA, AA, EPA. Dysregulated PUFA metabolism	[18]
18 advanced rectal cancer patients: CAPOX-treatment (plasma)	MRM-LC-MS/MS	LPE (22:5/0:0), SM (d18:2/18:1), LPC (16:0/0:0), LPC (15:1(9z)/0:0) and PC (40:2) are lower in NRP	[33]
20 myeloid leukemia patients (plasma)	(QTOF)UPLC-ESI-MS. GC-MS	Increase of AA precursors in leukemia patients' plasma. New targets for drug therapy	[5]
30 nascent MetS patients (plasma)	(QTOF) LC-MS/MS	Increases of PC (34:2) in patients with MetS. Novel biomarker in MetS.	[34]

Pre-analysis	Instrumentation	Analysis/Post-analysis	Ref.
Disease (biological sample)	**(Mass analyzer) system**	**Results/Conclusions**	
172 AD patients (plasma)	(LTQ) UHPLC-ESI-MS/MS	TAG (50:1), DAG (18:1) and PE (36:2) present in brain atrophy.	[35]
15 DM-1 and diabetic nephropathy children (urine)	(Triple TOF) UHPLC-MS	Increase of Cer (44;0,2) and HexCer species, suggesting as biomarkers of renal function decline.	[36]
70 endometriosis (plasma/peritoneal)	(QTOF) LC-MS/MS	Biomarkers: LPC 16:0, PE O-20:0, PE O 34:1, PC 36:2, PC 36:4, PC 36:5, PC 38:4, PC 38:6 and SM 34:1.	[37]
Liver/gastric/lung/colorectal/thyroid cancer (plasma-urine)	(ion trap) nUHPLC-ESI-MS/MS	LPE and PS high in thyroid cancer. Validation of cancer-specific lipid markers	[38]
432 DM-1 patients (plasma)	(triple Q)LC-MS/MS	lactosyl-Cer: predicts macroalbuminuria in DM-1	[39]
Multiple sclerosis patients (plasma)	(triple TOF) LC-MS	Cer-induced DNA-methylation of antiproliferative genes.	[40]
67 unipolar/bipolar disorders patients (plasma)	(Q Trap) LC-ESI-MS/MS	Increases: Cer (C16, C18, C20, C22, C24, C24:1, C24:1GluCer, C24 lactosylceramide), DAG, TAG	[41]
20 colorectal cancer patients (colon tissue)	LC-MS/MS	Increases of LPC, LPE, LPI (18:1) and LPI (18:0)	[42]
63 cutaneous leishmaniosis patients: treatment (plasma)	(triple TOF) LC-MS/MS	LTB4, 5-HETE, 5-oxo-HETE, 12-HETE, 11-HETE, PGE2, and 15-HETE. Targets of therapy	[43]
29 SARS-CoV-2 patients (plasma)	UHPLC-MS/MS	Alterations in PAs, sterols, SPHs and LPAs. Increases of Cer-phosphorylethanolamine and PE.	[44]
13 multiple sclerosis patients (post mortem brain tissue)	(Q-Trap) LC-MS/MS	Multiple sclerosis lesions: decrease: dhCer, Cer and SM subspecies. Increase: HexCer, Cer 1-phosphate	[45]
47 mTLE-HS patients (hippocampal sclerosis)	(QTOF) UPLC-ESI- -MS	33 lipids expressed. Decreased: Cer and lactosylceramide levels in mTLE-HS patients.	[46]
40 MetS patient (plasma)	(QTOF)/(Q-orbitrap) UHPLC-MS	PC: 18:1/16:0, o-22:3/22:3, P-18:1/16:1. Choline metabolism is affected. Biomarkers with prediction.	[24]
221 myopia children/adolescents (serum)	UHPLC-MS	275 metabolite presents in 33 pathways	[47]
106 colorectal cancer patient (tumor tissue)	(Q-Trap) HRMS/LC- MS/MS	Presence: LPC (16:1, 18:1, 20:4, 22:6) and SM species. Cer: C24:0–C26:0. GPL, GL and SM	[48]
Mild/moderate/severe asthma patients (bronchoscopy)	(orbitrap)-UPLC-MS	Increase: PC, LPC, bis (monoacylglycerol) phosphate. Decrease: OXPHOS (severe asthma).	[49]
33 lupus erythematosus patients (blood)	(triple Q) LC-MS/MS	Increase: LPL, PS species. Decrease of plasmalogen.	[50]

Pre-analysis	Instrumentation	Analysis/Post-analysis	Ref.
Disease (biological sample)	**(Mass analyzer) system**	**Results/Conclusions**	
826 AD cirrhosis patients (serum)	LC-MS/MS	SM permit distinguish between patients with compensated and decompensated cirrhosis.	[51]
15 Niemann-Pick disease patients (plasma)	LC-MS/MS	Increase: DG, AA and CE.	[52]
COVID-19 asymptomatic patients (serum)	(TOF) UHPLC-TIMS	15 lipids present in asymptomatic COVID-19 patients.	[53]
29 lung adenocarcinoma (tumor tissue)	(Q-orbitrap) LC-MS/MS	Increases of free-cholesterol and CE (18:1 and 20:4)	[54]
37 persons recovered/severe (SARSCoV-2) (plasma donor)	(QTOF) HILIC-LC-MS	Levels of fatty acyls and GPL were lower in recovered patients	[55]
SARS-CoV-2 peripheral leukocytes, colon/jejunum (plasma)	HPCL-MS/MS	Infection involving EPA, AA and gonadal steroids. ω-3 FFA associated with SARS-CoV-2 receptors	[56]
132 metastatic castration-resistant prostate cancer patients ENZA/AA drugs (plasma)	LC-MS	Increased Cer was associated with androgen receptor signaling inhibitors resistance.	[57]
60 CHD/HLP patients: *Salvia milthiorriza* treated (plasma)	(QTOF) UPLC-MS	Presence: PC (18:0/18:4; 18:2/16:0), PE (15:0 /22:1), LPC (0:0/18:0). *S. miltiorrhiza* reduce lipids	[58]
256 arthritis rheumatoid patients (serum)	(Q-orbitrap) UPLC-MS	Biomarkers: PE 16:0-18:2, TG 18:0-18:1-18:2	[13]
Patient: proteinuria, Fabry disease (plasma)	(Orbitrap) HRMS-UHPLC-MS	Galabiosylceramide-related lipid biomarker was higher in the patient's renal tissue biopsy	[12]
126 COVID-19 patients (serum)	(QTOF) UHPLC-MS	Biomarker: LPC 22:6, PC 36:1, bile acids. Lipidomics/machine learning techniques	[59]
206 obstructive sleep apnea patients (plasma)	(QTOF) UHPLC-ESI-MS/MS	GPL and bile acids are present. Adaptive mechanisms in response to obstructive sleep apnea	[60]
20 radiation/atherosclerotic carotid plaques patients	(QTOF) DESI-UPLC-MS	Biomarkers: 6 TG in the radiation-induced carotid plaques and atherosclerotic carotid plaques.	[61]
112 cystic fibrosis patients: drugs ELX/TEZ/IVA (plasma)	LC-MS/MS	Decrease: Cer (C16, C18, C20, C24:1). Up-regulation dhCer C24. ELX/TEZ/IVA	[62]

*Spectroscopy method.

Table 1.
Methodology, instrumentation, and human diseases diagnosed by lipidomic profiles.

In CKD, TAG, and N–acyl taurines increased, although total lipid and cholesterol levels, usually evaluated in clinical biochemistry, remain unchanged [29]. In patients infected with DNefHIV, lipidomic profiles revealed differences in the abundance of PS and SPH [63]. ACLF patients present with a particular lipid profile, mainly comprising CEs. These lipids also predict AD treatment outcomes [51]. In adverse pregnancies, the STBM contained the potential biomarkers SM, Chol, PS, PC, and PI [27]. In patients with CVD, the presence of the Cer d18:1/16:0, Cer d18:1/24:1, and Cer d18:1/24:0 were associated with a fatal outcome [8]. In DM-1, lipidomics can predict changes in SPH distribution associated with increased vascular permeability in different organs; this occurs only during the early onset of the disease and plays a critical role in developing complications [39]. In mTLE-HS patients, alterations to the hippocampus lipidome with potential for lipidomics-based therapies were reported. [46]. Lysophosphatidyl ethanolamine and lysophosphatidyl inositol species were associated with a number of cancer types. They are present in the liver (four PI and DAG 16:1_18:0), gastric (PI 34:2, 36:3, and 36:4, and LPA 18:2), lung (LPI 16:0, SM d18:1/20:0 and TAG 50:1 and 54:4), and thyroid cancer (LPI 18:0 and 18:1) [38]. Lipid differences between tumor and normal tissue have been proved to be of diagnostic and therapeutic significance [64]. Meanwhile, the upregulation of SPH downregulation of PI and glycerol phospholipid metabolism are associated with worse survival in patients with adrenocortical carcinoma [65].

In patients with depressive conditions, oxylipin concentration fluctuates between depression states. Dietary ω-6 (LA, AA) and ω-3 (EPA and DHA) fatty acids may underlie inflammatory states in symptomatic major depressive disorder with a seasonal pattern [66]; while brain lipidome changes include decreased PI, PA, and CL contents have been described [67]. Obese children with NAFLD showed increased hepatic epoxyeicosanoids with higher grades of steatosis and unaltered PUFA precursors [68]. In addition, in ACLF, Lipoxin A5 and epoxy keto octadecenoic acid formed a signature associated with coagulation and liver failures [69]. Moreover, in myocardial infarction, lipidomics showed an association between 2-aminoadipic acid and alterations of plasma metabolic signaling of hexoses, amino acids, biogenic amines, acylcarnitines, glycerophospholipids, and SPH showing the diagnostic and prognostic limits in acute and chronic heart failure [70].

Exposure to environmental pollutants can also alter lipid profiles; Cer, SPH, and TAG are potential biomarkers of lipotoxicity [71]. Macrophages reprogram their lipid metabolism in response to environmental cues [2]. Finally, there is an association between ether lipid signature and exceptional human longevity [72]. The diversity of conditions associated with differences in lipid profiles supports the growing importance of lipidomics as a clinical diagnostic tool.

2.2.1 Lipidomics in animal health

The study of lipidomics focuses not only on human health issues but also on animal health topics. Lipidomics helped explain the dynamics of inflammation during a bacterial attack in bovine mastitis [73] and provided a diagnostic biomarker for fatty liver disease in dairy cows [74]. Similarly, lipidomics contributed to understanding cystic fibrosis lung disease in newborn pigs [75] and identifying ganglioside disease markers in cats undergoing gene therapy [76]. Examples like these are poised to increase in the coming years as lipidomics extends its prognostic threshold to humans and animals, both economically relevant and wild.

2.3 Lipidomics in the study of rare diseases

Rare disorders remain a challenge even to modern medicine, and mounting evidence shows the prominent role of lipidomics in this scenario [12]. In patients with

Gaucher disease, the lipid profile allowed the diagnosis during enzyme replacement therapy [28]. Olmsted syndrome caused striking decreases in 15-LOX and dhCer levels [77]. In patients with Fabry disease, the presence of glycosphingolipids, galabiosyl-ceramide, and globotriaosylsphingosine is observed in vascular endothelium, nerves, cardiomyocytes, and renal glomerular podocytes [12]. A dysregulation in lipids involved in both cellular structure and membrane integrity was identified in fragile X syndrome, suggesting that X chromosomal deletion disorders are not limited to alterations in neuronal functions [78]. Barth syndrome showed a significant decrease in linoleic acid (18:2)-enriched molecular species, most notably tetra-18:2 (18:2-18:2-18:2-18:2), which is the major molecular species of cardiolipin in the myocardium [79].

2.3.1 Tay-Sachs Disease: an unwanted inheritance

The Tay-Sachs disease (TD) is a very rare (1:300 000) autosomal-recessive lysosomal storage disease [80]. This disorder is caused by mutations in the HEXA gene, which encodes for the β-hexosaminidase (Hex) enzyme [81]. Hex deficiency or low activity causes fatal inherited disorders [80]. Lysosomal enzyme Hex degrades the GM2 ganglioside; gangliosides are glycolipids present in neuronal cell plasma membranes. There are different Hex isoenzymes. HexA has two subunits (α and β), which are encoded by HEXA and HEXB genes, respectively. Hex B is a homodimer (two β-subunits), and HexS is a homodimer (two α-subunits) (**Figure 1**) [82].

Lyso-GM2 ganglioside concentrations in plasma and the brain of TD patients are elevated in association with loss of alpha-hexosaminidase activity. This molecule is thus a useful diagnostic and monitoring biomarker for this disease [83].

2.4 Lipidomics of COVID-19

The ongoing COVID-19 pandemic caused by SARS-CoV-2 has demonstrated the devastating impact of a critical illness on a global scale [44, 84]. In an equally unprecedented effort, the scientific community joined efforts to understand this disease from every conceivable point of view. Almost expectedly, SARS-CoV-2 infection disrupts the cellular lipid profile. Serum from recovered patients displayed different cytokine and lipidomic compositions than patients with severe disease [55]. The overall lipid metabolism increased the production of short- and medium-chain production of saturated fatty acids, acyl-carnitines, and SPH [44]. Lipidomic signatures, including n-3 long-chain PUFA, hydroxy fatty acids, and female gonadal steroids, were linked to SARS-CoV-2 infection [56]. Moreover, in a study by Castane et al. (2022), O-octanoyl-R-carnitine, LPE, AA, and oxylipins were the most altered parameters in COVID-19 patients compared to healthy volunteers, although the number of cases studied was small [59]. SARS-CoV-2 dysregulates lipid metabolism, especially the enhanced membrane phospholipid synthesis, and alters SPH homeostasis, implicating the specific host immune, inflammatory, and antiviral responses in asymptomatic COVID-19 [53] (Hao et al., 2021). Global response to COVID-19 infection highlighted the importance of cost- and time-effective biomarker detection, which is a requirement that lipidomic and machine learning fulfilled in a timely manner [59].

2.5 Lipidomics of pharmacologic interactions

Despite constant scientific advances, resistance and adverse reactions remain a pressing issue in clinical practice [85]; several groups report lipid profile alterations as

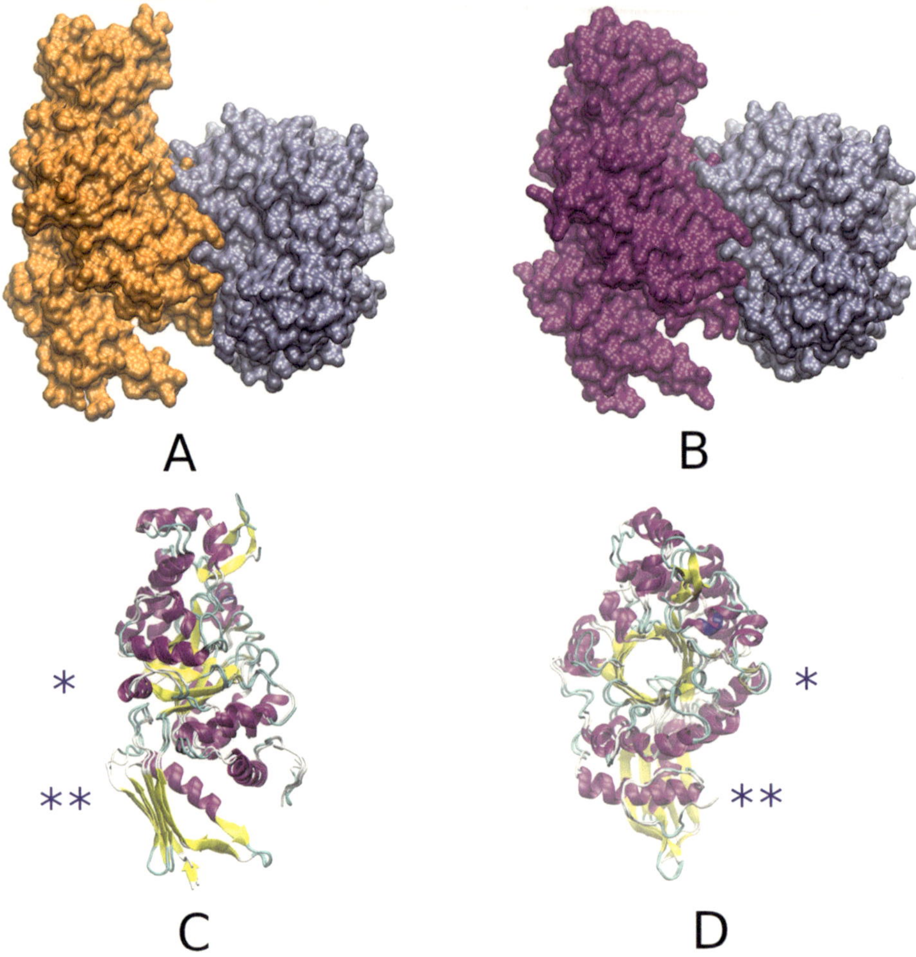

Figure 1.
Human β-hexosaminidase. A) Human β-hexosaminidase isoenzyme A structure (PDB ID: 2GJX) comprises two subunits from different genes. The subunit α (orange surface) is encoded by the HEXA gene (UniProtKB: P06865), and subunit β (blue light surface) is encoded by HEXB (UniProtKB: P07686). B) Hex isoenzyme A structure (PDB ID: 1NOU) comprises two β subunits (purple and blue light surfaces). C) Secondary structure representation of α and β subunits superimposed (helix in purple, β strands arrows in yellow, and turns/coil in cyan/white, respectively) in the same position that subunits orange and purple on A and B. Two CATH domains form both subunits; on the N-terminal, an α-β 2-layer sandwich architecture (**), followed the C-terminal by a catalytic α/beta-barrel fold architecture (*), the typical circular beta-barrel from the TPI enzymes. D) Secondary structure representation of α and β subunits superimposed rotated to see the α/β-barrel fold (*).

reactions to treatment. The antitumoral drug Imidazole Ketone Erastin increased the DAG, monoacylglycerol, and phospholipids, possibly through activation of the TAG hydrolysis enzyme (ATGL) in response to oxidative stress [86]. The CAPOX drugs employed in colorectal cancer treatment elicited differential levels of SM (d18:2/18:1), LysoPC (16:0/0:0), LysoPC (15:1(9z)/0:0), and Lyso PE (22:5/0:0) in responders versus non-responders [33]. In paclitaxel-resistant breast cancer, the forkhead box transcription factor M1 increased TG and PC, and decreased phospholipase D1 and lipid droplets [87]. In brain tumors, GPD1 displayed specific expression in brain tumor stem cells [88]. In general, lipid metabolism irradiation therapy (RT) disrupts the regulation of lipogenic genes, decreasing LPCs and cholesterol [89]. Meanwhile, in patients with distal esophageal cancer, RT induced cardiotoxicity, detecting six metabolites, after a four-week RT therapy [17].

Similar effects have been observed in human cancer-derived cell lines. Oxaliplatin, a frequent first-line adjuvant therapy for colorectal cancer, altered TAG and phospholipid levels in HT29 cells [85]. Diclofenac altered the cell phospholipid metabolism and induced PUFA accumulation in a neuroblastoma-derived cell line, suggesting tumoricidal potential [90]. The experimental antitumoral drug T-3764518 induced lipidomic changes and suppressed PC desaturation indices in HCT-116 cells [10]. Another experimental drug, FTY720, induced elevated levels of sphingosine, causing apoptosis in leukemic natural killer cells [91]. When treated with inducers of sodium phenylbutyrate (SPB) and all-trans retinoic acid (ATRA), the glioblastoma-derived cell line U87-MG displayed an increase in saturated PCs (38:1), 816 m/z; PC (36:1), 788 m/z; (31:1), 725 m/z, and a decrease in saturated PCs (PC (32:0), 734 m/z). These modifications in the lipidomic profile have potential application in therapy personalization [64].

Not only cancer cells respond to treatment with lipid metabolism imbalances. The VEN/OL drugs increased Cer (C18, C22C, and C24) in patients with depression or bipolar disorder [41]. Statin therapy to treat atherogenic dyslipidemia induced LPC and LPI [92]. Astaxanthin treatment revealed the over-accumulation of myocardial Cer in cardiac fibrosis [93]. Leishmaniasis biomarkers predicted by machine learning and lipidomic profile, such as eotaxin, 11-HETE, and transforming growth factor-β, were useful in identifying potential treatment failure [43]. Older HIV+ Australian men on antiretroviral therapy displayed high lipid dysregulation, specifically in GM3 ganglioside and mon ohexosylceramides, previously identified as frailty biomarkers [94]. Eicosanoid concentration was an indicator of other lipidic alterations in asthmatic subjects with aspirin intolerance [95]. Obese patients with insulin-resistant hypertriglyceridemic hypertension treated with statins showed increased plasmalogen and PUFA levels and reduced PE and PG classes [96]. In obstructive sleep apnea patients, a five-lipid group comprising 25-cinnamoyl-vulgaroside, glycocholic acid, bilirubin, and two previously unreported lipid species changed significantly after continuous positive airway pressure treatment [60]. Canagliflozin treatment increased the amounts of prostaglandin E2 and resolvin E3 in the liver of obese mice used as a biological model to understand the NAFLD [97]. Saroglitazar, a PPAR α/γ agonist, protected patients against obesity, insulin resistance, and steatosis by reducing TG and modulating phospholipid levels. Meanwhile, Hepano, an Ayurveda formulation, did so by modulating phospholipids, Cer, and oxidized lipids [98]. The ELX/TEZ/IVA modulator therapy alters plasma SPH levels and Cer species in cystic fibrosis patients [62]. Systemic lupus erythematosus patients treated with antioxidants displayed an ordered lipid conformation that contrasted with that of untreated patients [50]. The experimental drug J147 used to treat AD, reduced plasma FFA levels [99]. Donepezil, an anti-dementia drug, and the traditional Chinese medicine herbal decoction prepared to treat AD caused modifications in 15 types of compounds derived from PC, SM, and LPC, which are now considered potential lipid biomarkers [100].

As noted in these works, the lipidome is highly sensitive to a wide range of stimuli, so there is vast potential for developing drugs that selectively target these modifications [2]. As Wolf and collaborators (2008) note, lipidomic tools offer a practical option for diagnosis and treatment monitoring in many diseases [101].

2.6 Lipidomics in response to diet: we are what we eat

Recent literature shows a growing interest in lipidome modifications derived from the diet. Pomegranate seed oil and bitter melon extract modify the lipidomic profile of the cardiotoxicity induced by the anti-cancerous therapies showing anti-carcinogenic or cardioprotective properties [102].

The effect of maternal diet supplementation with conjugated linoleic acids influenced the contents of micro-elements in the cardiac tissue of newborns significantly [103]. Orange juice-derived nanovesicles modified the lipidome, decreasing TAG levels [104]. Tangeretin, a flavonoid present in some fruits, reduced body weight gain and ameliorated hepatic steatosis, lowering FFA, DG, TAG, Cer, and Chol levels, as detected by hepatic lipidomic analysis [105]. The diet supplemented with Lactococcus lactis (subsp. cremoris) increased glucose tolerance, developed less liver fat and inflammation, and decreased the oxylipin levels [106]. Moreover, in humans, phytosterol- and ω-3-supplemented milk reduced the LDL-GPL and LPC, inducing cardioprotection in persons with dyslipidemia and metabolic risk [107]. Salmon consumption induced selective incorporation of n-3 PUFA into PC lipids in human plasma, reducing cardiovascular disease risk [108]. Finally, herbal decoctions as in the traditional Chinese [109] and Asian [110] medicine resulted in an improved lipidomic profile in several human

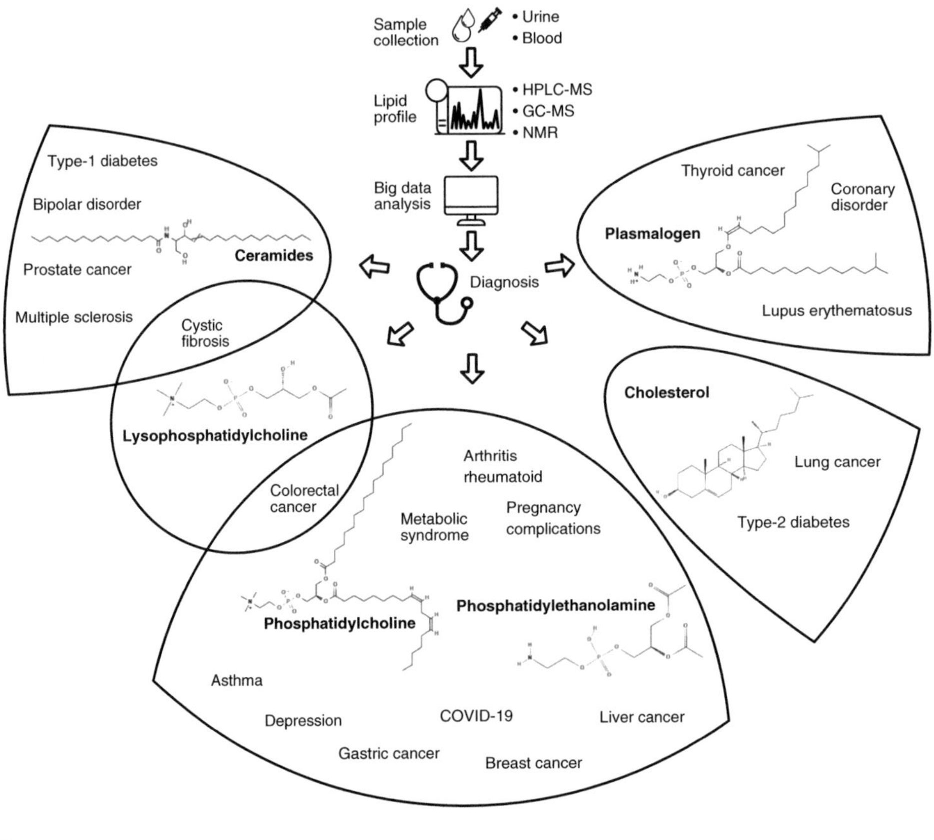

Figure 2.
The biological sample is obtained by noninvasive techniques such as blood and urine samples. Subsequently, the sample is analyzed by spectroscopic (NMR) and/or (mass spectrometry) techniques. Next, the results are compared with spectroscopic and/or spectrometric libraries. Soon, it will be possible to recognize the patients' lipid profile focused on the diagnosis or pharmacological treatment.

diseases. Together, these findings show that diet's therapeutic effects on the lipidome are evident in various human diseases. Ultimately, a clinical-chemical-bioinformatics method is proposed, using recurrent chemical techniques such as NMR [111] and mass spectrometry [112] , for the study of the lipidome in various human diseases (**Figure 2**).

3. Conclusions and perspectives

We live well into the era of big data and the discovery of biomarkers in various diseases through omics tools, such as lipidomics, which are currently in progress. Omics data is becoming increasingly essential for the diagnostic study and monitoring of human diseases and even in animals of economic importance. Therefore, lipidomics is suggested as a "gold standard" technique for the clinical and therapeutic part of this new era in clinical medicine. Finally, we agree with Hyötyläinen and collaborators (2017) [113], who reported that the lipidomic test must become inexpensive, and its added value concerning health economics needs to be demonstrated in a prospective setting.

Conflict of interest

The authors declare no conflict of interest.

Abbreviations

PC	phosphatidylcholine
LPC	lysophosphatidylcholine.
NAFLD	Nonalcoholic fatty liver disease.
DM	diabetes mellitus.
STBM	placental syncytiotrophoblast microvesicles.
PG	phosphatidylglycerol.
GPL	glycerophospholipids.
PI	phosphatidylinositol.
PS	phosphatidylserine.
PE	phosphatidylethanolamine.
SM	sphingomyelin.
Chol	cholesterol.
LPE	lysophosphatidylethanolamine.
LPI	lysophosphatidylinositol.
HDL-C	high-density lipoprotein-cholesterol.
DAG	diacylglycerol.
TAG	triacylglycerol.
FFA	free fatty acids.
FC	free cholesterol.
GPD1	glycerol-3-phosphate dehydrogenase.
CKD	chronic kidney disease.
ACLF	acute on chronic liver failure
CVD	cardiovascular disease.
AA	arachidonic acid.

EPA	eicosapentaenoic acid.
DHA	docosahexaenoic acid.
PUFA	polyunsaturated fatty acids.
LA	linoleic acid.
ALA	alpha-linolenic acid.
CAPOX	capecitabine/oxaliplatin treatment.
VEN/OL	venlafaxine and olanzapine treatment.
ENZA/AA	enzalutamide/abiraterone treatment.
MetS	metabolic syndrome.
15-LOX	15-lipoxygenase.
Cer	ceramide.
dhCer	dihydroceramide.
HexCer	hexoseceramide.
Hex	β-hexosaminidase.
GM2	disialotetrahexosylganglioside 2.
PPAR	peroxisome proliferator-activated receptors.
AD	Alzheimer's disease.
TD	Tay-Sachs disease.
ATGL	adipose triglyceride lipase.
NRP	not responder.
HETE	hydroxy eicosatetraenoic acid.
SARS-CoV-2	Severe acute respiratory syndrome coronavirus 2 of the genus Betacoronavirus.
SPH	sphingolipids.
LPA	lysophosphatidic acid.
mTLE-HS	mesial temporal lobe epilepsy-hippocampal sclerosis.
CE	cholesteryl ester.
CHD	Coronary heart disease.
ELX/TEZ/IVA	elexacaftor/tezacaftor/ivacaftor therapy.
GC-MS	gas chromatography coupled with mass spectrometer.
NMR H^1	proton nuclear magnetic resonance.
Q-trap	linear trap mass spectrometer.
TLC	thin layer chromatography.
HILIC	hydrophilic interaction liquid chromatography.
ESI	electrospray ionization.
MS/MS	tandem mass spectrometry.
LTQ	linear trap quadrupole mass spectrometer.
SIM	selected ion monitoring.
QTOF	quadrupole time-of-flight.
MRM	Multiple Reaction Monitoring.
Triple Q	triple quadrupole.
Q-orbitrap	quadrupole-orbiting trap.
HRMS	high-resolution mass spectrometer.
DESI	desorption electrospray ionization.
TIMS	thermal ionization mass spectrometry.
HPLC	high-performance liquid chromatography.
UHPLC	Ultra-HPLC.
nUHPLC	nano-HPLC.

Author details

María Elizbeth Alvarez Sánchez[1], Erick Nolasco Ontiveros[2], Rodrigo Arreola[3], Adriana Montserrat Espinosa González[2], Ana María García Bores[2], Roberto Eduardo López Urrutia[4], Ignacio Peñalosa Castro[2], María del Socorro Sánchez Correa[5] and Edgar Antonio Estrella Parra[2*]

1 Genomic Sciences Postgraduate, Autonomous University of Mexico City (UACM), México City, México

2 Faculty of Higher Studies-Iztacala (FES-I), Laboratory of Phytochemistry, UBIPRO, National Autonomous University of Mexico (UNAM), México City, México

3 Psychiatric Genetics Department, Clinical Research Branch, National Institute of Psychiatry, México City, México

4 Faculty of Higher Studies-Iztacala (FES-I), Laboratory of Functional Genomic, UBIMED, National Autonomous University of Mexico (UNAM), State of México, México

5 Faculty of Higher Studies (FES), Laboratory of Functional Genomic of Legumes, National Autonomous University of Mexico (UNAM), State of México, México

*Address all correspondence to: estreparr@iztacala.unam.mx

References

[1] Quehenberger O, Armando AM, Brown AH, Milne SB, Myers DS, Merrill AH, et al. Lipidomics reveals a remarkable diversity of lipids in human plasma. Journal of Lipid Research. 2010;**51**(11):3299-3305. DOI: 10.1194/jlr.M009449

[2] Hsieh WY, Zhou QD, York AG, Williams KJ, Scumpia PO, Kronenberger EB, et al. Toll-like receptors induce signal-specific reprogramming of the macrophage lipidome. Cell Metabolism. 2020;**32**(1):128-143. DOI: 10.1016/j.cmet.2020.05.003

[3] Roy J, Dibaeinia P, Fan TM, Sinha S, Das A. Global analysis of osteosarcoma lipidomes reveal altered lipid profiles in metastatic versus nonmetastatic cells. Journal of Lipid Research. 2019;**60**(2):375-387. DOI: 10.1194/jlr.M088559

[4] Yan SK, Liu RH, Jin HZ, Liu XR, Ye J, Shan L, et al. "Omics" in pharmaceutical research: Overview, applications, challenges, and future perspectives. Chinese Journal of Natural Medicines. 2015;**13**(1):3-21. DOI: 10.1016/S1875-5364(15)60002-4

[5] Pabst T, Kortz L, Fiedler GM, Ceglarek U, Idle JR, Beyoğlu D. The plasma lipidome in acute myeloid leukemia at diagnosis in relation to clinical disease features. BBA Clinical. 2017;**7**:105-114. DOI: 10.1016/j.bbacli.2017.03.002

[6] Slatter DA, Aldrovandi M, O'Connor A, Allen SM, Brasher CJ, Murphy RC, et al. Mapping the human platelet lipidome reveals cytosolic phospholipase A2 as a regulator of mitochondrial bioenergetics during activation. Cell Metabolism. 2016;**23**(5):930-944. DOI: 10.1016/j.cmet.2016.04.001

[7] Muralidharan S, Shimobayashi M, Ji S, Burla B, Hall MN, Wenk MR, et al. A reference map of sphingolipids in murine tissues. Cell Reports. 2021;**35**(11):109250. DOI: 10.1016/j.celrep.2021.109250

[8] Schmidt S, Gallego SF, Zelnik ID, Kovalchuk S, Albæk N, Sprenger RR, et al. Silencing of ceramide synthase 2 in hepatocytes modulates plasma ceramide biomarkers predictive of cardiovascular death. Molecular Therapy. 2022;**30**(4):1661-1674. DOI: 10.1016/j.ymthe.2021.08.021

[9] DeVeaux SA, Ogle ME, Vyshnya S, Chiappa NF, Leitmann B, Rudy R, et al. Characterizing human mesenchymal stromal cells' immune-modulatory potency using targeted lipidomic profiling of sphingolipids. Cytotherapy. 2022;**2022**(22):S1465-S3249. DOI: 10.1016/j.jcyt.2021.12.009

[10] Nishizawa S, Sumi H, Satoh Y, Yamamoto Y, Kitazawa S, Honda K, et al. In vitro and in vivo antitumor activities of T-3764518, a novel and orally available small molecule stearoyl-CoA desaturase 1 inhibitor. European Journal of Pharmacology. 2017;**807**:21-31. DOI: 10.1016/j.ejphar.2017.03.064

[11] Shi ZQ, Wang LY, Zheng JY, Xin GZ, Chen L. Lipidomics characterization of the mechanism of Cynomorium songaricum polysaccharide on treating type 2 diabetes. Journal of Chromatography. B, Analytical Technologies in the Biomedical and Life Sciences. 2021;**1176**:122737. DOI: 10.1016/j.jchromb.2021.122737

[12] Yazd HS, Bazargani SF, Vanbeek CA, King-Morris K, Heldermon C, Segal MS, et al. LC-MS lipidomics of renal biopsies for the diagnosis of Fabry disease.

Journal of Mass Spectrometry Advanced Clinical Laboratory. 2021a;**22**:71-78. DOI: 10.1016/j.jmsacl.2021.11.004

[13] Zhou GL, Jiawei X, Tingting L, Yan C, Wenjun W, Jue K, et al. Clinical lipidomics analysis reveals biomarkers of lipid peroxidation in serum from patients with rheumatoid arthritis. Microchemical Journal. 2021;**169**:106607. DOI: 10.1016/j.microc.2021.106607

[14] Abdullah L, Evans JE, Montague H, Reed JM, Moser A, Crynen G, et al. Chronic elevation of phosphocholine containing lipids in mice exposed to Gulf War agents pyridostigmine bromide and permethrin. Neurotoxicology and Teratology. 2013;**40**:74-84. DOI: 10.1016/j.ntt.2013.10.002

[15] Abdullah L, Evans JE, Joshi U, Crynen G, Reed J, Mouzon B, et al. Translational potential of long-term decreases in mitochondrial lipids in a mouse model of Gulf War Illness. Toxicology. 2016;**372**:22-33. DOI: 10.1016/j.tox.2016.10.012

[16] Begum H, Tortac F, Narayanaswamy P, Mundra PA, Ji S, Bendt AK, et al. Lipidomic profiling of plasma in a healthy Singaporean population to identify ethnic specific differences in lipid levels and associations with disease risk factors. Clinical Mass Spectrometry. 2017;**6**:25-31. DOI: 10.1016/j.clinms.2017.11.002

[17] Unger K, Li Y, Yeh C, Barac A, Srichai MB, Ballew EA, et al. Plasma metabolite biomarkers predictive of radiation induced cardiotoxicity. Radiotherapy and Oncology. 2020;**152**:133-145. DOI: 10.1016/j.radonc.2020.04.018

[18] Shih PB. Integrating multi-omics biomarkers and postprandial metabolism to develop personalized treatment for anorexia nervosa. Prostaglandins & Other Lipid Mediators. 2017;**132**:69-76. DOI: 10.1016/j.prostaglandins.2017.02.002

[19] Duarte-Delgado NP, Cala MP, Barreto A, Rodríguez CLS. Metabolites and metabolic pathways associated with rheumatoid arthritis and systemic lupus erythematosus. Journal of Translational Autoimmunity. 2022;**5**:100150. DOI: 10.1016/j.jtauto.2022.100150

[20] Paapstel K, Kals J, Eha J, Tootsi K, Ottas A, Piir A, et al. Inverse relations of serum phosphatidylcholines and lysophosphatidylcholines with vascular damage and heart rate in patients with atherosclerosis. Nutrition, Metabolism, and Cardiovascular Diseases. 2018;**28**(1):44-52. DOI: 10.1016/j.numecd.2017.07.011

[21] Hosseinkhani S, Aazami H, Hashemi E, Dehghanbanadaki H, Adibi-Motlagh B, Razi F. The trend in application of omics in type 2 diabetes researches; A bibliometric study. Diabetes and Metabolic Syndrome: Clinical Research and Reviews. 2021;**15**(5):102250

[22] Wang H, Zhang L, Zhang X, Song J, Guo Q, Zhang X, et al. Prediction model for different progressions of Atherosclerosis in ApoE-/- mice based on lipidomics. Journal of Pharmaceutical and Biomedical Analysis. 2022;**214**:114734. DOI: 10.1016/j.jpba.2022.114734

[23] Voros S, Maurovich-Horvat P, Marvasty IB, Bansal AT, Barnes MR, Vazquez G, et al. Precision phenotyping, panomics, and system-level bioinformatics to delineate complex biologies of atherosclerosis: Rationale and design of the "Genetic Loci and the Burden of Atherosclerotic Lesions" study. Journal of Cardiovascular Computed Tomography. 2014;**8**(6):442-451. DOI: 10.1016/j.jcct.2014.08.006

[24] Gong LL, Yang S, Zhang W, Han FF, Lv YL, Xuan LL, et al. Discovery of metabolite profiles of metabolic syndrome using untargeted and targeted LC-MS based lipidomics approach. Journal of Pharmaceutical and Biomedical Analysis. 2020;**177**:112848. DOI: 10.1016/j.jpba.2019.112848

[25] Ollero M, Astarita G, Guerrera IC, Sermet-Gaudelus I, Trudel S, Piomelli D, et al. Plasma lipidomics reveals potential prognostic signatures within a cohort of cystic fibrosis patients. Journal of Lipid Research. 2011;**52**(5):1011-1022. DOI: 10.1194/jlr.P013722

[26] Brindisi MC, Guiu B, Duvillard L, Athias A, Rollot F, Bouillet B, et al. Liver fat content is associated with an increase in cholesterol synthesis independent of statin therapy use in patients with type 2 diabetes. Atherosclerosis. 2012;**224**(2):465-468. DOI: 10.1016/j.atherosclerosis.2012.08.016

[27] Baig S, Lim JY, Fernandis AZ, Wenk MR, Kale A, Su LL, et al. Lipidomic analysis of human placental syncytiotrophoblast microvesicles in adverse pregnancy outcomes. Placenta. 2013;**34**(5):436-442. DOI: 10.1016/j.placenta.2013.02.004

[28] Byeon SK, Lee JY, Lee JS, Moon MH. Lipidomic profiling of plasma and urine from patients with Gaucher disease during enzyme replacement therapy by nanoflow liquid chromatography-tandem mass spectrometry. Journal of Chromatography. A. 2015;**1381**:132-139. DOI: 10.1016/j.chroma.2015.01.004

[29] Reis A, Rudnitskaya A, Chariyavilaskul P, Dhaun N, Melville V, Goddard J, et al. Top-down lipidomics of low density lipoprotein reveal altered lipid profiles in advanced chronic kidney disease. Journal of Lipid Research. 2015;**56**(2):413-422. DOI: 10.1194/jlr.M055624

[30] Schmöcker C, Kassner U, Kiesler S, Bismarck M, Rothe M, Steinhagen-Thiessen E, et al. A lipidomic analysis approach in patients undergoing lipoprotein apheresis. Atherosclerosis. 2016;**249**:30-35. DOI: 10.1016/j.atherosclerosis.2016.03.019

[31] Hadj Ahmed S, Koubaa N, Kharroubi W, Zarrouk A, Mnari A, Batbout F, et al. Identification of long and very long chain fatty acids, plasmalogen-C16:0 and phytanic acid as new lipid biomarkers in Tunisian coronary artery disease patients. Prostaglandins & Other Lipid Mediators. 2017;**131**:49-58. DOI: 10.1016/j.prostaglandins.2017.08.001

[32] Zhang R, Zhou Q, Cai X, Dong S, Le Z, Cai X, et al. Lipidomic analysis reveals the significant increase in diacyl glycerol phosphocholines in umbilical cord blood from pregnant women with gestational hypercholesterolemia. Placenta. 2017;**59**:39-45. DOI: 10.1016/j.placenta.2017.08.004

[33] Del Boccio P, Perrotti F, Rossi C, Cicalini I, Di Santo S, Zucchelli M, et al. Serum lipidomic study reveals potential early biomarkers for predicting response to chemoradiation therapy in advanced rectal cancer: A pilot study. Advances in Radiation Oncology. 2017;**2**(2):118-124. DOI: 10.1016/j.adro.2016.12.005

[34] Ramakrishanan N, Denna T, Devaraj S, Adams-Huet B, Jialal I. Exploratory lipidomics in patients with nascent Metabolic Syndrome. Journal of Diabetes and its Complications. 2018;**32**(8):791-794. DOI: 10.1016/j.jdiacomp.2018.05.014

[35] Kim SH, Yang JS, Lee JC, Lee JY, Lee JY, Kim E, et al. Lipidomic alterations in lipoproteins of patients with mild cognitive impairment and Alzheimer's disease by asymmetrical flow field-flow fractionation and nanoflow ultrahigh

performance liquid chromatography-tandem mass spectrometry. Journal of Chromatography. A. 2018;**1568**:91-100. DOI: 10.1016/j.chroma.2018.07.018

[36] Magagnotti C, Zerbini G, Fermo I, Carletti RM, Bonfanti R, Vallone F, et al. Identification of nephropathy predictors in urine from children with a recent diagnosis of type 1 diabetes. Journal of Proteomics. 2019;**193**:205-216. DOI: 10.1016/j.jprot.2018.10.010

[37] Starodubtseva N, Chagovets V, Borisova A, Salimova D, Aleksandrova N, Chingin K, et al. Identification of potential endometriosis biomarkers in peritoneal fluid and blood plasma via shotgun lipidomics. Clinical Mass Spectrometry. 2019;**13**:21-26. DOI: 10.1016/j.clinms.2019.05.007

[38] Lee GB, Lee JC, Moon MH. Plasma lipid profile comparison of five different cancers by nanoflow ultrahigh performance liquid chromatography-tandem mass spectrometry. Analytica Chimica Acta. 2019;**1063**:117-126. DOI: 10.1016/j.aca.2019.02.021

[39] Lopes-Virella MF, Baker NL, Hunt KJ, Hammad SM, Arthur J, Virella G, et al. Glycosylated sphingolipids and progression to kidney dysfunction in type 1 diabetes. Journal of Clinical Lipidology. 2019;**13**(3):481-491. DOI: 10.1016/j.jacl.2019.03.005

[40] Castro K, Ntranos A, Amatruda M, Petracca M, Kosa P, Chen EY, et al. Body mass index in multiple sclerosis modulates ceramide-induced DNA methylation and disease course. eBioMedicine. 2019;**43**:392-410

[41] Brunkhorst-Kanaan N, Klatt-Schreiner K, Hackel J, Schröter K, Trautmann S, Hahnefeld L, et al. Targeted lipidomics reveal derangement of ceramides in major depression and bipolar disorder. Metabolism. 2019;**95**:65-76. DOI: 10.1016/j.metabol.2019.04.002

[42] Wang Y, Hinz S, Uckermann O, Hönscheid P, von Schönfels W, Burmeister G, et al. Shotgun lipidomics-based characterization of the landscape of lipid metabolism in colorectal cancer. Biochimica et Biophysica Acta - Molecular and Cell Biology of Lipids. 2020;**1865**(3):158579. DOI: 10.1016/j.bbalip.2019.158579

[43] Malta-Santos H, Fukutani KF, Sorgi CA, Queiroz ATL, Nardini V, Silva J, et al. Multi-omic analyses of plasma cytokines, lipidomics and transcriptomics distinguish treatment outcomes in cutaneous Leishmaniasis. iScience. 2020;**23**(12):101840. DOI: 10.1016/j.isci.2020.101840

[44] Thomas T, Stefanoni D, Dzieciatkowska M, Issaian A, Nemkov T, Hill RC, et al. Evidence of structural protein damage and membrane lipid remodeling in red blood cells from COVID-19 patients. Journal of Proteome Research. 2020;**19**(11):4455-4469. DOI: 10.1021/acs.jproteome.0c00606

[45] Podbielska M, Szulc ZM, Ariga T, Pokryszko-Dragan A, Fortuna W, Bilinska M, et al. Distinctive sphingolipid patterns in chronic multiple sclerosis lesions. Journal of Lipid Research. 2020;**61**(11):1464-1479. DOI: 10.1194/jlr.RA120001022

[46] Zhang H, Ren P, Huang Y, Zeng W, Zhong K, Gao H, et al. Untargeted lipidomic analysis of human hippocampus for temporal lobe epilepsy with hippocampal sclerosis. Epilepsy Research. 2020;**161**:106299. DOI: 10.1016/j.eplepsyres.2020.106299

[47] Du B, Jin N, Zhu X, Lu D, Jin C, Li Z, et al. A prospective study of serum metabolomic and lipidomic changes

in myopic children and adolescents. Experimental Eye Research. 2020;**199**:108182. DOI: 10.1016/j.exer.2020.108182

[48] Ecker J, Benedetti E, Kindt ASD, Höring M, Perl M, Machmüller AC, et al. The colorectal cancer lipidome: Identification of a robust tumor-specific lipid species signature. Gastroenterology. 2021;**161**(3):910-923. DOI: 10.1053/j.gastro.2021.05.009

[49] Ravi A, Goorsenberg AWM, Dijkhuis A, Dierdorp BS, Dekker T, van Weeghel M, et al. Metabolic differences between bronchial epithelium from healthy individuals and patients with asthma and the effect of bronchial thermoplasty. The Journal of Allergy and Clinical Immunology. 2021;**148**(5):1236-1248. DOI: 10.1016/j.jaci.2020.12.653

[50] Hu C, Zhang J, Hong S, Li H, Lu L, Xie G, et al. Oxidative stress-induced aberrant lipid metabolism is an important causal factor for dysfunction of immunocytes from patients with systemic lupus erythematosus. Free Radical Biology & Medicine. 2021;**163**:210-219. DOI: 10.1016/j.freeradbiomed.2020.12.006

[51] Clària J, Curto A, Moreau R, Colsch B, López-Vicario C, Lozano JJ, et al. Untargeted lipidomics uncovers lipid signatures that distinguish severe from moderate forms of acutely decompensated cirrhosis. Journal of Hepatology. 2021;**75**(5):1116-1127. DOI: 10.1016/j.jhep.2021.06.043

[52] Boenzi S, Catesini G, Sacchetti E, Tagliaferri F, Dionisi-Vici C, Deodato F. Comprehensive-targeted lipidomic analysis in Niemann-Pick C disease. Molecular Genetics and Metabolism. 2021;**34**(4):337-343. DOI: 10.1016/j.ymgme.2021.11.005

[53] Hao Y, Zhang Z, Feng G, Chen M, Wan Q, Lin J, et al. Distinct lipid metabolic dysregulation in asymptomatic COVID-19. iScience. 2021;**24**(9):102974. DOI: 10.1016/j.isci.2021.102974

[54] Hoppstädter J, Dembek A, Höring M, Schymik HS, Dahlem C, Sultan A, et al. Dysregulation of cholesterol homeostasis in human lung cancer tissue and tumour-associated macrophages. eBioMedicine. 2021;**72**:103578. DOI: 10.1016/j.ebiom.2021.103578

[55] Acosta-Ampudia Y, Monsalve DM, Rojas M, Rodríguez Y, Gallo JE, Salazar-Uribe JC, et al. COVID-19 convalescent plasma composition and immunological effects in severe patients. Journal of Autoimmunity. 2021;**18**:102598. DOI: 10.1016/j.jaut.2021.102598

[56] Mayneris-Perxachs J, Moreno-Navarrete JM, Ballanti M, Monteleone G, Alessandro Paoluzi O, Mingrone G, et al. Lipidomics and metabolomics signatures of SARS-CoV-2 mediators/receptors in peripheral leukocytes, jejunum and colon. Computational and Structural Biotechnology Journal. 2021;**19**:6080-6089. DOI: 10.1016/j.csbj.2021.11.007

[57] Lin HM, Mak B, Yeung N, Huynh K, Meikle TG, Mellett NA, et al. Overcoming enzalutamide resistance in metastatic prostate cancer by targeting sphingosine kinase. eBioMedicine. 2021;**72**:103625. DOI: 10.1016/j.ebiom.2021.103625

[58] Sun G, Li X, Wei J, Zhang T, Li B, Chen M, et al. Pharmacodynamic substances in Salvia miltiorrhiza for prevention and treatment of hyperlipidemia and coronary heart disease based on lipidomics technology and network pharmacology analysis. Biomedicine & Pharmacotherapy. 2021;**141**:111846. DOI: 10.1016/j.biopha.2021.111846

[59] Castane H, Iftimie S, Baiges-Gaya G, Rodríguez-Tomàs E, Jiménez-Franco A, et al. Machine learning and semi-targeted lipidomics identify distinct serum lipid signatures in hospitalized COVID-19-positive and COVID-19-negative patients. Metabolism. 2022;**131**:155197. DOI: 10.1016/j.metabol.2022.155197

[60] Pinilla L, Benítez ID, Santamaria-Martos F, Targa A, Moncusí-Moix A, Dalmases M, et al. Plasma profiling reveals a blood-based metabolic fingerprint of obstructive sleep apnea. Biomedicine & Pharmacotherapy. 2022;**145**:112425. DOI: 10.1016/j.biopha.2021.112425

[61] Li W, Wang T, Zhang X, Zhu J, Li XY, Peng F, et al. Distinct lipid profiles of radiation-induced carotid plaques from atherosclerotic carotid plaques revealed by UPLC-QTOF-MS and DESI-MSI. Radiotherapy and Oncology. 2022;**167**:25-33. DOI: 10.1016/j.radonc.2021.12.006

[62] Westhölter D, Schumacher F, Wülfinghoff N, Sutharsan S, Strassburg S, Kleuser B, et al. CFTR modulator therapy alters plasma sphingolipid profiles in people with cystic fibrosis. Journal of Cystic Fibrosis. 2022;**2022**(22):S1569. DOI: 10.1016/j.jcf.2022.02.005

[63] Low H, Cheng L, Di Yacovo MS, Churchill MJ, Meikle P, Bukrinsky M, et al. Lipid metabolism in patients infected with Nef-deficient HIV-1 strain. Atherosclerosis. 2016;**244**:22-28. DOI: 10.1016/j.atherosclerosis.2015.10.103

[64] Liu H, Wang S, Lin JM, Lin Z, Li HF. Investigation of the lipidomic changes in differentiated glioblastoma cells after drug treatment using MALDI-MS. Talanta. 2021;**233**:122570. DOI: 10.1016/j.talanta.2021.122570

[65] Subramanian C, Cohen MS. Identification of novel lipid metabolic biomarkers associated with poor adrenocortical carcinoma prognosis using integrated bioinformatics. Surgery. 2022;**171**(1):119-129. DOI: 10.1016/j.surg.2021.04.049

[66] Hennebelle M, Otoki Y, Yang J, Hammock BD, Levitt AJ, Taha AY, et al. Altered soluble epoxide hydrolase-derived oxylipins in patients with seasonal major depression: An exploratory study. Psychiatry Research. 2017;**252**:94-101. DOI: 10.1016/j.psychres.2017.02.056

[67] Faria R, Santana MM, Aveleira CA, Simões C, Maciel E, Melo T, et al. Alterations in phospholipidomic profile in the brain of mouse model of depression induced by chronic unpredictable stress. Neuroscience. 2014;**273**:1-11. DOI: 10.1016/j.neuroscience.2014.04.042

[68] Kalveram L, Schunck WH, Rothe M, Rudolph B, Loddenkemper C, Holzhütter HG, et al. Regulation of the cytochrome P450 epoxyeicosanoid pathway is associated with distinct histologic features in pediatric non-alcoholic fatty liver disease. Prostaglandins, Leukotrienes, and Essential Fatty Acids. 2021;**164**:102229. DOI: 10.1016/j.plefa.2020.102229

[69] López-Vicario C, Checa A, Urdangarin A, Aguilar F, Alcaraz-Quiles J, Caraceni P, et al. Targeted lipidomics reveals extensive changes in circulating lipid mediators in patients with acutely decompensated cirrhosis. Journal of Hepatology. 2020;**73**(4):817-828. DOI: 10.1016/j.jhep.2020.03.046

[70] Halade GV, Kain V, Tourki B, Jadapalli JK. Lipoxygenase drives lipidomic and metabolic reprogramming in ischemic heart failure. Metabolism. 2019;**96**:22-32. DOI: 10.1016/j.metabol.2019.04.011

[71] Li R, Wang Y, Hou B, Lam SM, Zhang W, Chen R, et al. Lipidomics

insight into chronic exposure to ambient air pollution in mice. Environmental Pollution. 2020;**262**:114668. DOI: 10.1016/j.envpol.2020.114668

[72] Pradas I, Jové M, Huynh K, Puig J, Ingles M, Borras C, et al. Exceptional human longevity is associated with a specific plasma phenotype of ether lipids. Redox Biology. 2019;**21**:101127. DOI: 10.1016/j.redox.2019.101127

[73] Ryman VE, Pighetti GM, Lippolis JD, Gandy JC, Applegate CM, et al. Quantification of bovine oxylipids during intramammary Streptococcus uberis infection. Prostaglandins & Other Lipid Mediators. 2015;**121** (Pt B):207-217. DOI: 10.1016/j.prostaglandins.2015.09.006

[74] Gerspach C, Imhasly S, Gubler M, Naegeli H, Ruetten M, Laczko E. Altered plasma lipidome profile of dairy cows with fatty liver disease. Research in Veterinary Science. 2017;**110**:47-59. DOI: 10.1016/j.rvsc.2016.10.001

[75] Bréa D, Soler L, Fleurot I, Melo S, Chevaleyre C, Berri M, et al. Intrinsic alterations in peripheral neutrophils from cystic fibrosis newborn piglets. Journal of Cystic Fibrosis. 2020;**19**(5):830-836. DOI: 10.1016/j.jcf.2020.02.016

[76] Gray-Edwards HL, Jiang X, Randle AN, Taylor AR, Voss TL, Johnson AK, et al. Lipidomic evaluation of feline neurologic disease after AAV gene therapy. Molecular Therapies Methods Clinical Development. 2017;**6**:135-142. DOI: 10.1016/j.omtm.2017.07.005

[77] Wakabayashi M, Yoshioka T, Higashino K, Numata Y, Igarashi Y, Kihara A. Decreases in 15-lipoxygenase metabolites in Olmsted syndrome model rats. Journal of Dermatological Science. 2017;**85**(3):186-196. DOI: 10.1016/j.jdermsci.2016.12.013

[78] Yazd HS, Rubio VY, Chamberlain CA, Yost RA, Garrett TJ. Metabolomic and lipidomic characterization of an X-chromosome deletion disorder in neural progenitor cells by UHPLC-HRMS. Journal of Mass Spectrometry Advanced Clinical Laboratory. 2021b;**20**:11-24. DOI: 10.1016/j.jmsacl.2021.05.002

[79] Kiebish MA, Yang K, Liu X, Mancuso DJ, Guan S, Zhao Z, et al. Dysfunctional cardiac mitochondrial bioenergetic, lipidomic, and signaling in a murine model of Barth syndrome. Journal of Lipid Research. 2013;**54**(5):1312-1325. DOI: 10.1194/jlr.M034728

[80] Lefter S, Mahony O, Sweeney B, Ryan AM. Late-onset Tay-Sachs disease in an irish family. Movement Disorder Clinical Practise. 2021;**8**(1):106-110. DOI: 10.1002/mdc3.13096

[81] Vu M, Li R, Baskfield A, Lu B, Farkhondeh A, Gorshkov K, et al. Neural stem cells for disease modeling and evaluation of therapeutics for Tay-Sachs disease. Orphanet Journal of Rare Diseases. 2018;**13**(1):152. DOI: 10.1186/s13023-018-0886-3

[82] Cavender C, Mangini L, Van Vleet JL, Corado C, McCullagh E, Gray-Edwards HL, et al. Natural history study of glycan accumulation in large animal models of GM2 gangliosidoses. PLoS One. 2020;**15**(12):e0243006. DOI:10.1371/journal.pone.0243006

[83] Kodama T, Togawa T, Tsukimura T, Kawashima I, Matsuoka K, Kitakaze K, et al. Lyso-GM2 ganglioside: A possible biomarker of Tay-Sachs disease and Sandhoff disease. PLoS One. 2011;**6**(12):e29074. DOI: 10.1371/journal.pone.0029074

[84] McKenna HT, O'Brien KA, Fernandez BO, Minnion M, Tod A, McNally BD, et al. Divergent trajectories

of cellular bioenergetics, intermediary metabolism and systemic redox status in survivors and non-survivors of critical illness. Redox Biology. 2021;**41**:101907. DOI: 10.1016/j.redox.2021.101907

[85] Yu J, Hu D, Cheng Y, Guo J, Wang Y, Tan Z, et al. Lipidomics and transcriptomics analyses of altered lipid species and pathways in oxaliplatin-treated colorectal cancer cells. Journal of Pharmaceutical and Biomedical Analysis. 2021;**200**:114077. DOI: 10.1016/j.jpba.2021.114077

[86] Zhang Y, Tan H, Daniels JD, Zandkarimi F, Liu H, Brown LM, et al. Imidazole ketone erastin induces ferroptosis and slows tumor growth in a mouse lymphoma model. Cell Chemical Biology. 2019;**26**(5):623-633.e9. DOI: 10.1016/j.chembiol.2019.01.008

[87] Zhang X, Huang C, Yuan Y, Jin S, Zhao J, Zhang W, et al. FOXM1-mediated activation of phospholipase D1 promotes lipid droplet accumulation and reduces ROS to support paclitaxel resistance in metastatic cancer cells. Free Radical Biology & Medicine. 2022;**179**:213-228. DOI: 10.1016/j.freeradbiomed.2021.11.024

[88] Rusu P, Shao C, Neuerburg A, Acikgöz AA, Wu Y, Zou P, et al. GPD1 specifically marks dormant glioma stem cells with a distinct metabolic profile. Cell Stem Cell. 2019;**25**(2):241-257. DOI: 10.1016/j.stem.2019.06.004

[89] Costa S, Fairfield H, Farrell M, Murphy CS, Soucy A, Vary C, et al. Sclerostin antibody increases trabecular bone and bone mechanical properties by increasing osteoblast activity damaged by whole-body irradiation in mice. Bone. 2021;**147**:115918. DOI: 10.1016/j.bone.2021.115918

[90] Hunt AN, Macken M, Koster G, Kohler JA, Postle AD. Diclofenac mediated derangement of neuroblastoma cell lipidomic profiles is accompanied by increased phosphatidylcholine biosynthesis. Advances in Enzyme Regulation. 2008;**48**:74-87. DOI: 10.1016/j.advenzreg.2007.11.005

[91] Liao A, Broeg K, Fox T, Tan SF, Watters R, Shah MV, et al. Therapeutic efficacy of FTY720 in a rat model of NK-cell leukemia. Blood. 2011;**118**(10):2793-2800. DOI: 10.1182/blood-2011-01-331447

[92] Chapman MJ, Orsoni A, Tan R, Mellett NA, Nguyen A, Robillard P, et al. LDL subclass lipidomics in atherogenic dyslipidemia: Effect of statin therapy on bioactive lipids and dense LDL. Journal of Lipid Research. 2020;**61**(6):911-932. DOI: 10.1194/jlr.P119000543

[93] Shi Y, Lin P, Wang X, Zou G, Li K. Sphingomyelin phosphodiesterase 1 (SMPD1) mediates the attenuation of myocardial infarction-induced cardiac fibrosis by astaxanthin. Biochemical and Biophysical Research Communications. 2018;**503**(2):637-643. DOI: 10.1016/j.bbrc.2018.06.054

[94] Yeoh HL, Cheng AC, Cherry CL, Weir JM, Meikle PJ, Hoy JF, et al. Immunometabolic and lipidomic markers associated with the frailty index and quality of life in aging HIV+ men on antiretroviral therapy. eBioMedicine. 2017;**22**:112-121. DOI: 10.1016/j.ebiom.2017.07.015

[95] Sanak M, Gielicz A, Bochenek G, Kaszuba M, Niżankowska-Mogilnicka E, Szczeklik A. Targeted eicosanoid lipidomics of exhaled breath condensate provide a distinct pattern in the aspirin-intolerant asthma phenotype. The Journal of Allergy and Clinical Immunology. 2011;**127**(5):1141-1147. DOI: 10.1016/j.jaci.2010.12.1108

[96] Meikle PJ, Wong G, Tan R, Giral P, Robillard P, Orsoni A, et al. Statin action favors normalization of the plasma lipidome in the atherogenic mixed dyslipidemia of MetS: Potential relevance to statin-associated dysglycemia. Journal of Lipid Research. 2015;**56**(12):2381-2392. DOI: 10.1194/jlr.P061143

[97] Yoshino K, Hosooka T, Shinohara M, Aoki C, Hosokawa Y, Imamori M, et al. Canagliflozin ameliorates hepatic fat deposition in obese diabetic mice: Role of prostaglandin E2. Biochemical and Biophysical Research Communications. 2021;**557**:62-68. DOI: 10.1016/j.bbrc.2021.04.012

[98] Sarkar S, Kumari D, Gupta SK, Sharma V, Mukhi S, et al. Saroglitazar and Hepano treatment offers protection against high fat high fructose diet induced obesity, insulin resistance and steatosis by modulating various class of hepatic and circulating lipids. Biomedicine & Pharmacotherapy. 2021;**144**:112357. DOI: 10.1016/j.biopha.2021.112357

[99] Kepchia D, Huang L, Currais A, Liang Z, Fischer W, Maher P. The Alzheimer's disease drug candidate J147 decreases blood plasma fatty acid levels via modulation of AMPK/ACC1 signaling in the liver. Biomedicine & Pharmacotherapy. 2022;**147**:112648. DOI: 10.1016/j.biopha.2022.112648

[100] Zhao NN, Sun Y-F, Zong L, Liu S, Song F-R, Liu Z-Q, et al. Serum lipidomics study of Ding-Zhi-Xiao-Wan effect on Alzheimer's disease using online liquid extraction surface analysis coupled to direct infusion mass spectrometry. International Journal of Mass Spectrometry. 2018;**434**:29-36

[101] Wolf C, Quinn PJ. Lipidomics in diagnosis of lipidoses. Sub-Cellular Biochemistry. 2008;**49**:567-588. DOI: 10.1007/978-1-4020-8831-5_22

[102] Bialek A, Bialek M, Lepionka T, Pachniewicz P, Czauderna M. Oxysterols and lipidomic profile of myocardium of rats supplemented with pomegranate seed oil and/or bitter melon aqueous extract - cardio-oncological animal model research. Chemistry and Physics of Lipids. 2021;**235**:105057. DOI: 10.1016/j.chemphyslip.2021.105057

[103] Bialek M, Bialek A, Ruszczyńska A, Bulska E, Zaworski K, Czauderna M. Evaluation of the influence of diet supplementation with conjugated linoleic acid isomers on elemental composition in the cardio-oncological nutritional programming rat' model. Journal of Trace Elements in Medicine and Biology. 2021;**68**:126816. DOI: 10.1016/j.jtemb.2021.126816

[104] Berger E, Colosetti P, Jalabert A, Meugnier E, Wiklander OPB, Jouhet J, et al. Use of nanovesicles from orange juice to reverse diet-induced gut modifications in diet-induced obese mice. Molecular Theraphy Methods in Clinical Development. 2020;**18**:880-892. DOI: 10.1016/j.omtm.2020.08.009

[105] Feng K, Lan Y, Zhu X, Li J, Chen T, Huang Q, et al. Hepatic lipidomics analysis reveals the antiobesity and cholesterol-lowering effects of tangeretin in high-fat diet-fed rats. Journal of Agricultural and Food Chemistry. 2020;**68**(22):6142-6153. DOI: 10.1021/acs.jafc.0c01778

[106] Naudin CR, Maner-Smith K, Owens JA, Wynn GM, Robinson BS, Matthews JD, et al. Lactococcus lactis Subspecies cremoris Elicits Protection against metabolic changes induced by a western-style diet. Gastroenterology. 2020;**159**(2):639-651. DOI: 10.1053/j.gastro.2020.03.010

[107] Padro T, Vilahur G, Sánchez-Hernández J, Hernández M,

Antonijoan RM, Perez A, et al. Lipidomic changes of LDL in overweight and moderately hypercholesterolemic subjects taking phytosterol- and omega-3-supplemented milk. Journal of Lipid Research. 2015;**56**(5):1043-1056. DOI: 10.1194/jlr.P052217

[108] Žáček P, Bukowski M, Johnson L, Raatz SK, Picklo M. Selective enrichment of n-3 fatty acids in human plasma lipid motifs following intake of marine fish. The Journal of Nutritional Biochemistry. 2018;**54**:57-65. DOI: 10.1016/j.jnutbio.2017.11.002

[109] Chao Y, Gao S, Li N, Zhao H, Qian Y, Zha H, et al. Lipidomics reveals the therapeutic effects of EtOAc extract of Orthosiphon stamineus Benth on nephrolithiasis. Frontiers in Pharmacology. 2020;**11**:1299. DOI: 10.3389/fphar.2020.01299

[110] Shon JC, Kim WC, Ryu R, Wu Z, Seo JS, Choi MS, et al. Plasma lipidomics reveals insights into anti-obesity effect of Chrysanthemum morifolium ramat leaves and its constituent luteolin in high-fat diet-induced dyslipidemic mice. Nutrients. 2020;**12**(10):2973. DOI: 10.3390/nu12102973

[111] Nodeland M, Klevjer M, Sæther J, Giskeødegård G, Bathen TF, Wisløff U, et al. Atherogenic lipidomics profile in healthy individuals with low cardiorespiratory fitness: The HUNT3 fitness study. Atherosclerosis. 2022;**343**:51-57. DOI: 10.1016/j.atherosclerosis.2022.01.001

[112] Ivanovová E, Piskláková B, Dobešová D, Kvasnička A, Friedecký D. Novel LC-MS tools for diagnosing inborn errors of metabolism. Microchemical Journal. 2021;**170**:106654. DOI: 10.1016/j.microc.2021.106654

[113] Hyötyläinen T, Ahonen L, Pöhö P, Orešič M. Lipidomics in biomedical research-practical considerations. Biochimica et Biophysica Acta - Molecular and Cell Biology of Lipids. 2017;**1862**(8):800-803. DOI: 10.1016/j.bbalip.2017.04.002